Springer Geophysics

The Springer Geophysics series seeks to publish a broad portfolio of scientific books, aiming at researchers, students, and everyone interested in geophysics. The series includes peer-reviewed monographs, edited volumes, textbooks, and conference proceedings. It covers the entire research area including, but not limited to, applied geophysics, computational geophysics, electrical and electromagnetic geophysics, geodesy, geodynamics, geomagnetism, gravity, lithosphere research, paleomagnetism, planetology, tectonophysics, thermal geophysics, and seismology.

More information about this series at http://www.springer.com/series/10173

S. P. Maurya · N. P. Singh · K. H. Singh

Seismic Inversion Methods: A Practical Approach

Springer

S. P. Maurya
Department of Geophysics
Institute of Sciences
Banaras Hindu University
Varanasi, Uttar Pradesh, India

N. P. Singh
Department of Geophysics
Institute of Sciences
Banaras Hindu University
Varanasi, Uttar Pradesh, India

K. H. Singh
Department of Earth Sciences
Indian Institute of Technology Bombay
Mumbai, Maharashtra, India

ISSN 2364-9119 ISSN 2364-9127 (electronic)
Springer Geophysics
ISBN 978-3-030-45661-0 ISBN 978-3-030-45662-7 (eBook)
https://doi.org/10.1007/978-3-030-45662-7

This Springer imprint is published by the registered company Springer Nature Switzerland AG
The registered company address is: Gewerbestrasse 11, 6330 Cham, Switzerland

Contents

Chapter 1
Fundamental of Seismic Inversion

Abstract Seismic inversion methods in geophysics is a technique used to transform seismic reflection data into quantitative subsurface rock properties. It is methods to integrate seismic reflection data along with well log data to extract a variety of petrophysical parameters. Geophysicists regularly perform seismic surveys to collect subsurface geological information in the exploration project. Such surveys record sound waves that have passed through the earth's layers of rock and fluid in the form of amplitude and time. The recorded seismic data can be interpreted on its own but this does not provide sufficient information of the subsurface and can be inaccurate under certain circumstances. Because of its efficiency and quality, most oil and gas companies now using seismic inversion methods to increase data resolution and reliability and improve rock properties estimation including porosity and net pay. In this chapter, the basics of forward modeling, convolution model, seismic inversions and their kinds are discussed.

1.1 Introduction

Seismic inversion is a procedure that helps extract underlying models of the physical characteristics of rocks and fluids from seismic and well-log data. In the absence of well data, the properties can also be inferred from the inversion of seismic data alone (Krebs et al. 2009). In the oil and gas industry, seismic inversion technique has been widely used as a tool to locate hydrocarbon-bearing strata in the subsurface from the seismic reflection data (Morozov and Ma 2009; Lindseth 1979). This method dramatically increases the resolution of seismic data and hence helps to interpret seismic data.

The physical parameters that are of interest to a modeler performing inversion are impedance (Z), P-wave (V_P) and S-wave (V_S) velocity and density (ρ). Lame parameters that are sensitive towards fluid and saturation in rocks (Clochard et al. 2009) can also be derived from inverted models of impedances. The petrophysical parameters like porosity, sand/shale ratio, gas saturation, etc. can be estimated further with the help of inverted volumes (Goodway 2001). These petrophysical parameters

S. P. Maurya et al., *Seismic Inversion Methods: A Practical Approach*, Springer Geophysics, https://doi.org/10.1007/978-3-030-45662-7_1

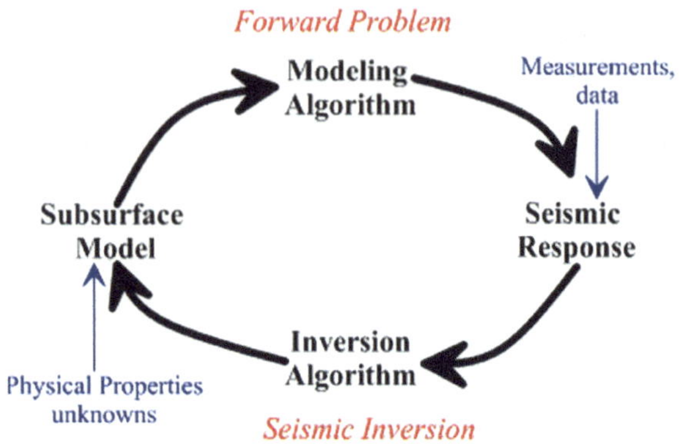

Fig. 1.1 A schematic diagram representing forward and inverse modeling processes

added strength to the seismic data interpretation which is a very crucial process for any exploration project.

To understand seismic inversion methods, one needs to first understand forward modeling. The seismic forward modeling uses the principle of convolution theory which states that seismic trace can be generated by the convolution of wavelet with earth reflectivity. The sound waves are sent to the subsurface that interacts with the earth's layer and returns on the surface which is recorded by the receivers. These recorded signals are called observations (Seismic signal) and the entire process is called forward modeling (Maurya et al. 2018). On the other hand, in seismic inversion methods, we have observation available and we seek to find the subsurface geological model that is a representation of observation. Both processes can be understood from Fig. 1.1.

1.2 Seismic Forward Modeling

There are many geophysical methods used to explore oil and gas from the subsurface but the most important technique is seismic imaging. The imaging means the visual representation of the earth's subsurface model. This is one of the ultimate goals of the geophysicist. Seismic forward modeling can be implemented by using both numerical studies in geoscience and computation technology. The forward modeling procedure uses an elastic impedance method that generates synthetic seismograms from velocities and densities of the subsurface layers (Connolly 1999). The elastic impedance at each interface is calculated as a function of the offset. The resulting impedance series is transformed into the reflectivity series and convolved with the source wavelet to get a stacked seismic gather. The impedance (Z) is computed from the product of velocity (v) and density ρ.

$$Z = V\rho \tag{1.1}$$

The zero offset reflectivity series can be calculated from the impedance as following.

$$R_j = \frac{Z_{j+1} - Z_j}{Z_{j+1} + Z_j} \tag{1.2}$$

where Z_j is the seismic impedance of jth layer, and R_j is seismic reflectivity of the interface between jth and $(j + 1)$th layer. The reflection coefficient for the angle-dependent incident wave is estimated using the following formula (Bachrach et al. 2014).

$$R(\theta) = \frac{1}{2}\left(1 - \tan^2\theta\right)\frac{\Delta I_p}{I_p} - 4\frac{V_s^2}{V_p^2}\sin^2\theta\frac{\Delta I_s}{I_s} - \left[\frac{1}{2}\tan^2\theta - 2\frac{V_s^2}{V_p^2}\sin^2\theta\right]\frac{\Delta\rho}{\rho} \tag{1.3}$$

The synthetic seismogram is calculated from the reflection coefficient using the following equation.

$$S(t) = W(t) * R(t) + N(t) \tag{1.4}$$

where $S(t)$ is synthetic seismogram, $W(t)$ is source wavelet, $R(t)$ is the reflection coefficient of the subsurface and $N(t)$ is additive noise and generally assumed to be zero for simplicity. The seismic forward modeling method gives the understanding of seismic travel time, elastic impedance, arrival time, earth's reflectivity, seismic amplitude generated by the seismic wave and the other aspects.

Figure 1.2 demonstrates the use of forward modeling in faulted and anticline geological models. Figure 1.2a shows geological model of the subsurface whereas Fig. 1.2b depicts the corresponding seismic section generated using the forward modeling technique. From the figure, we can see that the seismic gather (Fig. 1.2b) shows an approximately same geological structure which is in the geological model (Fig. 1.2a). Now our aim is to find this geological model from the observation i.e. seismic section which can be achieved by the seismic inversion methods.

1.3 Seismic Inversion

Seismic inversion methods involve mapping rock and fluid properties of the subsurface of the earth using seismic measurements made on the surface of the earth as input. In fact, all inversion methods aim to estimate the geophysical properties of the subsurface from the measurement made on the surface. In the seismic inversion process, there are three main issues that need to be addressed carefully to get a broadband spectrum with high-resolution images of the subsurface. The first problem arises due to the band-limited nature of the seismic data which means seismic

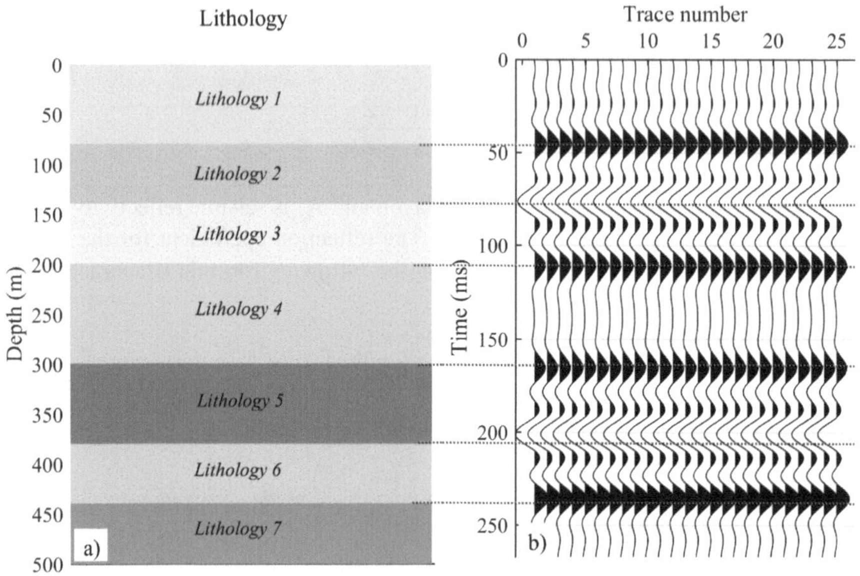

Fig. 1.2 **a** Shows geological model and **b** depicts seismic section expressed geological model

data does not have a low and high-frequency component. The seismic data generally have 10–80 Hz frequency and hence does not contains frequency less than 10 Hz and greater than 80 Hz however these frequencies are very important for the interpretation. On the other hand, well log data has both low and high frequencies and can be integrated with the seismic data to compensate for this. The second problem is to use a seismic wavelet. From the convolution theory, we have seen that the seismic wavelet is used to generate synthetic data and hence accurate wavelet estimation is critical for successful inversion results (Russell 1988; Maurya and Singh 2018). The third and most important problem is that the seismic inversion is non-unique. There may be possibly more than one solution to the same problem. To reduce the number of solution one needs other constraints to bind the inversion results. These constraints can be prior to geological information, well log data, etc.

Figure 1.3 shows an example of seismic inversion methods. Figure 1.3a shows seismic data from the Blackfoot field, Canada whereas Fig. 1.3b shows seismic inversion results. From Fig. (1.3) one can notice how dramatically the image quality has been increased. From the seismic section, one has an only amplitude and hence cannot interpret much. However from the inverted section, one can classify the sand formation, shale formation and hence can identify the productive zone.

To understand the seismic inversion technique, one must first understand the physical processes which are involved in generating these data. The seismic data is generated using a forward modeling technique that uses the convolution model for its implementation. Initially, therefore one should look at the basic convolutional model of the seismic trace in the time and frequency domains. This model has three

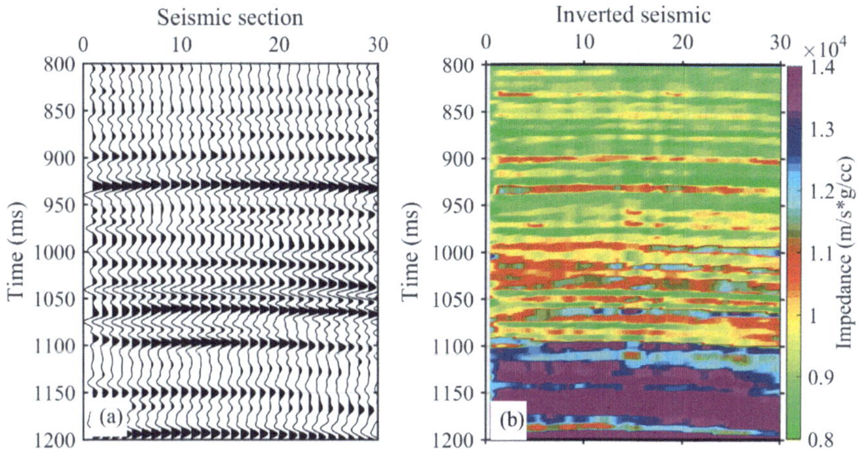

Fig. 1.3 **a** Depicts seismic section and **b** depicts inverted section

components, first is earth's reflectivity, second is seismic wavelet, and the third is the noise. After understanding the concepts and the problems which can occur, one is in a position to look at the methods which are currently used to invert seismic data.

1.4 The Convolution Model

The convolution model is the most common one-dimensional model for the seismic trace. The convolution model states that the seismic trace can be generated by the convolution of seismic wavelet with the earth's reflectivity series along with the addition of noises. It can be written mathematically as follows.

$$S(t) = R(t) * W(t) + N(t) \tag{1.5}$$

where * implies convolution process, $S(t)$ is a seismic trace, $R(t)$ is earth reflectivity, $W(t)$ is wavelet and $N(t)$ is the noise component. By considering perfect case one can consider noise component to be zero and hence Eq. 1.5 can be written simply as

$$S(t) = R(t) * W(t) \tag{1.6}$$

Equations 1.5 and 1.6 represents the convolution model in the time domain. An alternate form of the above equation is frequency domain, can be obtained by taking the Fourier transform of the above equations which results in

$$S(f) = W(f) \times R(f) \tag{1.7}$$

where $S(f)$ is seismic trace and presented in the frequency domain, $W(f)$ is wavelet in the frequency domain and $R(f)$ is the earth's reflectivity in the frequency domain.

From the above equation, we can see that the convolution in time domain becomes multiplication in the frequency domain. However, working in frequency domain is very complex, yet it is normal to consider the amplitude and phase spectra of each individual component (Russell 1988). The Eq. 1.7 can be simplified in the amplitude and phase spectrum as follows.

$$|S(f)| = |W(f)| \times |R(f)| \tag{1.8}$$

$$\theta_S(f) = \theta_W(f) + \theta_R(f) \tag{1.9}$$

where || indicates amplitude spectrum and θ indicates phase spectrum.

In other words, one can say that the convolution process involves multiplying the amplitude spectra of wavelet and reflectivity and adding their phase spectra individually. If one is able to suppress the noise component from the data, and then deconvolve with the wavelet give the earth's reflectivity series. This reflectivity series can be transformed into acoustic impedance which is the ultimate goal of any seismic inversion methods. Figure 1.4 shows the convolution process graphically. The first track of Fig. 1.4 shows reflectivity series generated randomly, track 2 and track 3 show minimum phase wavelet and seismic trace which are generated by the convolution of reflectivity with the source wavelet.

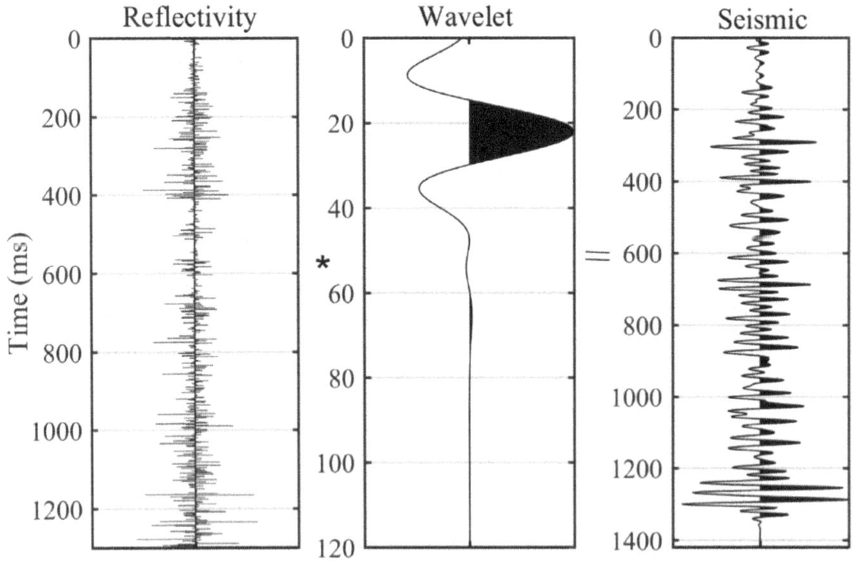

Fig. 1.4 Generation of seismic data by using the convolution model. Track 1 shows the earth's reflectivity, track 2 shows minimum phase wavelet and track 3 shows seismic trace

Now, after having a good understanding of the basic convolution model which is used by most of the seismic inversion methods. We are now in a position to look into the seismic inversion methods and their classifications. The seismic inversion methods are summarized in the following sections.

1.5 Classification of Seismic Inversion

The seismic inversion techniques can be divided into two broad categories, Post-stack, and Pre-stack inversion methods. The first approach is the most commonly used where the effect of the wavelet is removed from the seismic data and a high-resolution image of the subsurface is produced (Chen and Sidney 1997). The second approach relies on model building from well log, seismic and geological data (Downton 2005). This also generates a high-resolution image of the subsurface from which reservoir properties are calculated. A reliable estimate of the reservoir properties is critical in the decision-making process during the development phase (Pendrel 2006). These inversion methods are further divided into subparts which are discussed in the following sections.

1.5.1 Post-stack Seismic Inversion

Post-stack seismic inversion techniques are categorized as the first type. This type of inversion results in acoustic impedance volume utilizing seismic data through the integration of the well data and a basic stratigraphic interpretation. This impedance volume can be used to estimate reservoir properties away from well (Russell and Hampson 1991; Morozov and Ma 2009). Some of the advantages of post-stack inversion are mentioned below.

1. As the acoustic impedance is a layer property; hence stratigraphic interpretation is easier on impedance data than seismic data.
2. The reduction of wavelet effects, side lobes, and tuning enhance the resolution of subsurface layers.
3. Acoustic impedance can be directly computed and compared to well log measurements that serve as a link to reservoir properties.
4. Porosity can be related to the acoustic impedance. Using geostatistical methods these impedance volume can be transformed into the porosity volume within the reservoir.
5. Acoustic impedance can be utilized to locate individual reservoir regions.
6. It takes very less time than pre-stack inversion.
7. It does not give shear wave information to discriminate against the fluid effects.

Further, post-stack inversion can be divided into two parts namely deterministic inversion which includes model-based inversion methods and second is Stochastic inversion methods which include Band-limited inversion, Colored inversion, and sparse spike inversion methods. All these methods utilize post-stack seismic data and are inverted for the impedance section. All these methods have a different operating principle and they need proper understanding before applying to the data.

We will start by looking at the most common methods of post-stack inversion i.e. band-limited inversion methods. The Band-Limited recursive (iterative) inversion method was developed by Lindseth (1979). It is the most basic type of inversion which assumes that seismic amplitude is proportional to the earth reflectivities and transform the input seismic trace to acoustic impedance traces. The equation used by BLI is given below.

$$Z_{j+1} = Z_1 exp\left(\gamma \sum_{k=1}^{j} S_k\right) \tag{1.10}$$

where $S_k = 2r_k/\gamma$, Z is impedance and r is earth's reflectivity. It used to integrate the seismic trace and then exponentiates the result to provide an impedance trace. The input seismic data is usually wavelet processed. This does not however fully satisfy the basic assumptions since the wavelets are not completely removed from the data. Consequently, the tuning and the wavelet side lobe's effects remain. Moreover, the results are produced within the seismic bandwidth and errors in calculating the acoustic impedance (AI) for the subsequent layers get integrated (Maurya and Singh 2018; Maurya et al., 2019)

Thereafter, the second type of post-stack inversion methods i.e. model-based inversion is utilized. This method uses post-stack seismic data and computes acoustic impedance. The method is based on the convolutional theory which states that the seismic trace can be generated from the convolution of wavelet with the reflectivity function (Leite 2010; Maurya and Singh 2015a, b). This method model the subsurface as layers or blocks in terms of acoustic impedance and time. The initial impedance model is built from the interpolation of the impedance logs obtained from the wells in the area. The impedance of each layer may vary laterally and vertically. The impedance bonds are set to keep the optimized model laterally smooth within the given limits.

Another type of post-stack inversion technique used most commonly is the Colored inversion method. In this method, the inversion is represented as a convolutional process where an operator (O) in the frequency domain is used to transform the seismic traces (S) into impedance (Z): Z = O * S (Lancaster and Whitcombe 2000). This operator maps the seismic amplitude spectrum into the earth impedance spectrum. Spectra of Acoustic Impedance logs calculated from wells in the same area are used to derive the spectrum of the operator. The phase of this operator is 900 making it easy to integrate it with the reflectivity series to give impedance (Ansari 2014).

The sparse-spike method is another type of post-stack inversion methods used to estimate subsurface physical property. This method, unlike other techniques, gives an estimate of the reflectivity series that would approximate the seismic data with minimum number of (sparse) spikes. Non-uniqueness, in this case, is taken care of by applying the sparse reflectivity criterion. The maximum likelihood deconvolution and L_1 norms (Linear Programming) logarithms are used to achieve this (Banihasan et al. 2006).

1.5.2 Pre-stack Seismic Inversion

The Pre-stack inversion falls into the second category of the seismic inversion techniques. The estimate of the elastic properties of the subsurface such as the S-wave velocity of the subsurface layers which are sensitive to fluid saturation can be obtained from Pre-stack inversion (Moncayo et al. 2012).

Pre-stack inversion transforms seismic data (angle/offset gathers) into P-impedance, S-impedance and density volumes through the integration of well data and horizon information from seismic data. P-impedance and Vp/Vs ratio are reliable, depending on target depth and acquisition configuration, and can be used to predict reservoir properties away from well (Carrazzone et al. 1996). Pre-stack seismic inversion provides several benefits.

1. The P-impedance, S-impedance, and density give layer properties, whereas seismic data is an interface property.
2. Enhanced resolution of sub-surface layers due to the reduction of wavelet effects, tuning and side lobes.
3. Acoustic impedance can be directly compared to well log measurements which in turn are linked to reservoir properties.
4. Compared with other inversion techniques (e.g. post-stack inversion), the data offers additional information to distinguish between lithology and fluid effects.

The most common methods which fall in this category are simultaneous inversion, Elastic impedance inversion, and AVO inversion methods.

Simultaneous inversion is the first type of pre-stack seismic inversion. Pre-stack seismic gather contains additional information i.e. S-wave velocity which travels slowly in the subsurface and contains more information about the rock properties of the earth. This information can be estimated from the pre-stack gathers using several seismic inversion methods. A common approach is simultaneous inversion of pre-stack seismic data, which inverts for several rock property parameters simultaneously. The Aki and Richards (1980) formula gives access to the approximate reflectivity at the various offsets in the pre-stack domain and can be expressed as follows.

$$R(\theta) = a\frac{\Delta V_P}{V_P} + b\frac{\Delta\rho}{\rho} + c\frac{\Delta V_S}{V_S} \tag{1.11}$$

where $a = \frac{1}{2\cos^2\theta}$, $b = 0.5 - \left[2\left(\frac{V_S}{V_P}\right)^2 \sin^2\theta\right]$, $c = -4\left(\frac{V_S}{V_P}\right)^2 \sin^2\theta$,

$$\rho = \frac{\rho_1 + \rho_2}{2}, \Delta\rho = \rho_2 - \rho_1, V_P = \frac{V_{P1} + V_{P2}}{2},$$
$$\Delta V_P = V_{P2} - V_{P1}, V_S = \frac{V_{S1} + V_{S2}}{2},$$
$$\Delta V_S = V_{S2} - V_{S1}, \theta = \frac{\theta_1 + \theta_2}{2}$$

These functions are used to compute the earth's reflectivity. Thereafter, reflectivity is convolved with the wavelet to obtain the synthetic seismic gather. Further, this synthetic gather is compared with the original seismic gather of the area and the misfit is computed between them. The model is subsequently perturbed and a new comparison made to reduce the misfit and the least misfit model is selected as the final solution.

The other inversion method used commonly is the elastic impedance inversion. Connolly (1999) introduced the concept of Elastic Impedance and defined a function R which is dependent on the incidence angle and is related to the relative variation of Elastic Impedance.

$$R(\theta) \approx \frac{1}{2}\frac{\Delta EI}{EI} \approx \frac{1}{2}\Delta \ln(EI)$$

This function R is now called the Elastic Impedance, in analog to the acoustic impedance concept. The angle-dependent P-wave reflectivity is also approximated by the well-known simplified description of the Zoeppritz equations (Aki-Richards):

$$R(\theta) = A + B\sin^2\theta + C\tan^2\theta\sin^2\theta \tag{1.13}$$

where $A = \frac{1}{2}\left[\frac{\Delta_P}{V_P} + \frac{\Delta\rho}{\rho}\right]$, $B = \frac{1}{2}\frac{\Delta V_P}{V_P} - 4\left[\frac{V_S}{V_P}\right]^2\frac{\Delta V_S}{V_S} - 2\left[\frac{V_S}{V_P}\right]^2\frac{\Delta\rho}{\rho}$, $C = \frac{1}{2}\frac{\Delta V_P}{V_P}$.

Combining the two expressions, doing $k = [V_S/V_P]^2$ constant results in the Elastic Impedance being equal to:

$$EI = V_P^{(1+\tan^2\theta)} V_S^{(8K\sin^2\theta)} \rho^{(1-8K\sin^2\theta)} \tag{1.14}$$

Other types of pre-stack inversion techniques, Amplitude-Variation-with-Offset (AVO), have been widely used in hydrocarbon exploration over the previous two decades. Traditional AVO analysis involves calculating the term AVO intercept, gradient, and higher-order AVO from the fit of the amplitude of P-wave reflection to the sine of the square of angle of incidence. This model is focused, among others, on Bortfeld (1961) and Shuey (1985), the estimated P-wave reflection coefficient model in intercept-gradient shape. The AVO intercept and gradient values can also be coupled under the assumption of a background PS velocity ratio to acquire extra AVO characteristics such as the pseudo-S-wave information and Poisson's contrast ratio. In a hybrid inversion system, AVO intercept and pseudo-S-wave information are also used together with pre-stack waveform reversal (PSWI). Hybrid inversion is

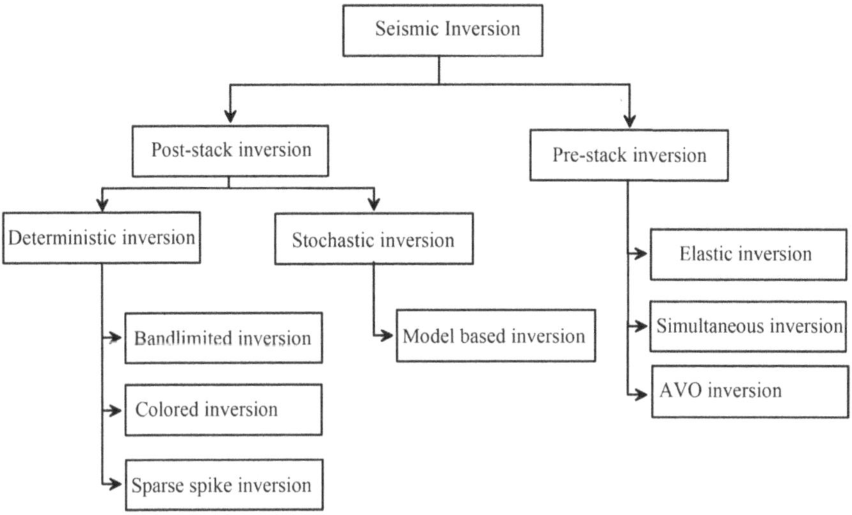

Fig. 1.5 Flowchart of classification of seismic inversion methods

a mixture of methodologies for pre-stack and post-stack inversion. This mix allows big data amounts to be efficiently reversed in the lack of good information.

Figure 1.5 discusses the classification of seismic inversion methods graphically although some recent development has been done which added more classification but above discussed inversion methods are convention techniques that are still in use to estimate subsurface properties from seismic and well log data. These inverted sections strengthen to seismic data interpretation and hence help to characterize the reservoir.

1.6 Local Optimization Methods

Local optimization method is a heuristic technique to solve computationally difficult problems of optimization. Local optimization can be used on problems that can be formulated as a solution that maximizes a criterion among a number of candidate solution alternatives. A typical local optimization scheme's flow chart is displayed in Fig. 1.6. Most local optimization systems are iterative algorithms, and the primary objective of all these algorithms is to ensure that the objective function is reduced at each iteration. These algorithms always try to move downhill and are therefore referred to as greedy algorithms (Xu et al. 2012). It is quite evident that the local algorithms in the neighborhood of the starting point are searching for a local minimum. The selection of a starting solution is therefore of paramount importance, a poor selection of the starting solution will result in the algorithm being trapped in a local minimum.

Fig. 1.6 Flow chart of local optimization methods

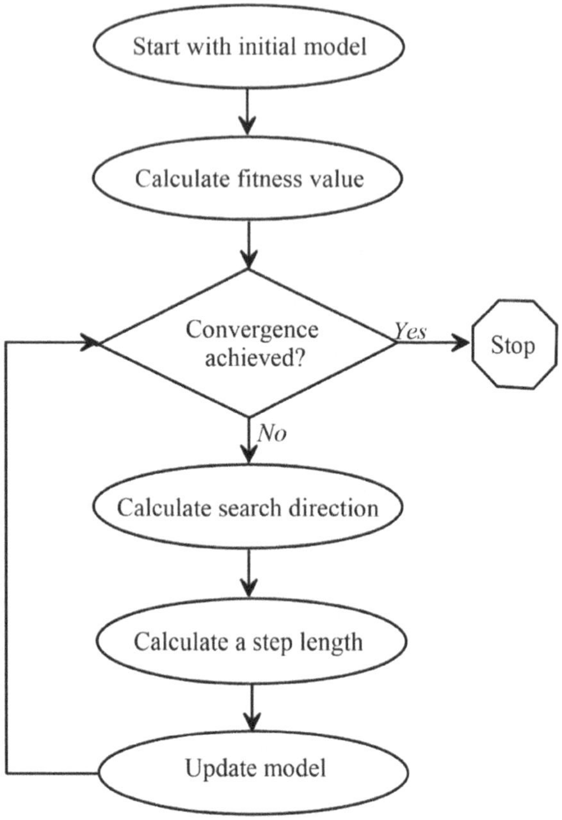

The algorithm calculates a search path using a local property of the objective function, determining an update or increase to the present model, given a starting solution. Only a fraction of the measured update is applied, however, and a step-length factor is used to ensure that the new objective function value is always lower than the current value (Chunduru et al. 1997). Thus the critical factors are the update calculations and the step length. Methods of optimization vary on the basis that how these two parameters are calculated.

1.7 Global Optimization Methods

Global methods of optimization are grooving techniques for methods of seismic inversion. Optimization involves finding the optimal value of a multi-variable function. The feature we want to minimize (or maximize) is a misfit (or fitness) function which characterizes the distinctions (or similarities) between observed and synthetic

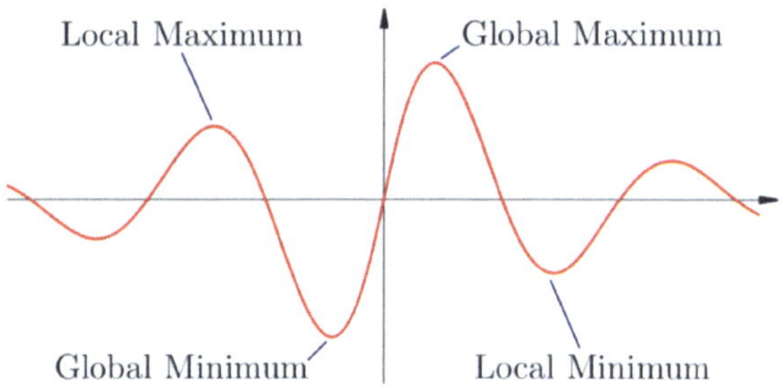

Fig. 1.7 Figure demonstrates the difference between local and global optimum

data calculated using an assumed earth model (Sen and Stoffa 2013). Physical parameters that characterize the characteristics of rock layers, such as compressional wave velocity, shear wave speed, resistivity, etc., describe the earth model.

The solution of a linear equation system is the same as a quadratic function minimization. Local minimization processes mentioned above are appropriate when there is a single minimum of the cost function. In particular, however, inverse geophysical problems are non-linear and extremely complicated. The cost function is probable to demonstrate multiple minima in such instances. In such cases, local optimization techniques are anticipated to fail as they tend to cover the closest local minima (Sen et al. 1995). Local minimization algorithms are anticipated to cover a fake solution and produce bad outcomes unless there is adequate a priori data to make an intelligent guess about the original model. Gradient-based optimization systems do not provide the means to jump from a local minimum to a worldwide solution (Maurya and Singh 2019). However, in most situations, computationally more expensive are the global optimization schemes compared to local optimization schemes. Figure 1.7 shows the difference between local and global solutions.

The primary motivation to develop efficient global optimization technique lies in the fact that the technique should be practicable in obtaining better results compared to the local optimization techniques in situations where the problem is complex and the model dimensionality is large. There are many global optimization methods available but Genetic algorithm and Simulated annealing are broadly used in the seismic inversion.

The technique of the genetic algorithm (GA) is based on the analogy that the genetic modifications that occur in the living species work towards making the species smarter and more adaptive to the changing natural environment. One of the powerful tools for global optimization is the genetic algorithm; it is also based on the principle of random walking in the space parameter (Sen and Oltz 2006; Maurya et al. 2018). GA has artificial intelligence that can handle issues that are highly nonlinear. GA needs no derivatives or information on curvature. Therefore, once the forward problem is solved, the reverse problem can be solved automatically because the whole exercise consists of selecting some models, calculating synthetic data, comparing d^{prs} with d^{obs}, calculating the cost function or error function and selecting better and better models through certain guidelines when deciding on the acceptance criterion (Sen and Stoffa 2013).

The simulated annealing (SA) uses techniques of random walking, but with some artificial intelligence. It brings the concept of temperature in this technique and which is controlling parameters even though the temperature has nothing to do with an inverse problem. The function of energy in thermodynamics is replaced by the function of error. This error function is also referred to in global optimization as cost function or energy function. In thermodynamics, the probability density functions of Gibbs can be expressed as follows.

$$P(E_j) = \frac{exp\left(\frac{-E_j}{KT}\right)}{\sum exp\left(\frac{-E_j}{KT}\right)} \tag{1.15}$$

where K is the constant of Boltzmann. K is assumed to be unity in global optimization because it has no role in iterative methods of inversion. Difference between the observed field data d^{obs} and the model's synthetic data, i.e. d^{pred} generates the error function, defined as follows.

$$E(m) = \left(d^{obs} - d^{pred}\right)^T \left(C_D^{-1}\right)\left(d^{obs} - d^{pred}\right) \tag{1.16}$$

where C_D is the data covariance. The inverse problem begins with a very high initial temperature of the controlling parameter. The temperature is progressively reduced in consecutive iteration to make Gibb's probability density function more and more sensitive (Vestergaard and Mosegaard 1991). In this manner, the cost function is minimized. This technique is discussed in detail in Chap. 6.

1.8 Geostatistical Inversion

Geostatistical methods are routinely followed to predict various geophysical parameters from seismic and well log data. The geostatistical methods use sample points taken at different locations and interpolate in the seismic section where log data are not available. These sample points are measurements of petrophysical parameters in the boreholes (Haas and Dubrule 1994). The geostatistics derives a surface using the values from the measured locations to estimate data points for each location in between the data points. Two groups of interpolation techniques are provided by Geostatistics namely deterministic and geostatistical (Russell et al. 1997). Mathematical functions are used in the deterministic techniques for interpolation whereas geostatistics uses both statistical and mathematical methods (Hampson et al. 2001).

There are four types of geostatistical method namely single attribute analysis, multi-attribute regression, Neural network which include multi-layer feed-forward and probabilistic neural network methods. The procedures of the geostatistical method are as follows.

1. The spatial continuity of the well log data is quantified using variograms.
2. A statistical relationship is derived between the log and seismic data at all well locations using cross-validation plots. The single attribute analysis, multivariate regression uses linear relationship whereas the Neural Network uses a non-linear relationship.
3. These linear and non-linear relationships are then used to estimate well log property away from the boreholes.
4. The predicted well log property is evaluated for its reliability.

In single attribute analysis, first, a variety of attributes are estimated directly or indirectly from the seismic data and are analyzed to get the best attribute. The best attribute selected on the basis of its correlation with the target log values. Further, the best attribute is cross plotted with target value from the well log data and a best fit straight line is chosen which gives the desired relationship. This desired relationship is then used to predict petrophysical parameters in the inter-well region.

Multi-attribute regression is the second type of geostatistical method used to predict petrophysical parameters. The basic principle is similar to the single attribute analysis. The multi-attribute analysis has only one difference from the single attribute analysis, in the sense that this method uses more than one attribute at a time. In this method, all the attributes are analyzed and the best combination of attributes (more than one) is selected and cross plotted with a target value to give linear relationship which is used for further analysis.

Till now the analysis is linear which have some limitation but now we are extending our analysis to nonlinear. The neural network falls in this category. It is divided into two parts namely multi-layer feed-forward neural network and probabilistic neural network. A multilayer feed-forward neural network is an interconnection of neurons in which data and calculations flow in a single direction, from the input data to the outputs with intermediate one or more hidden layers. Each layer consists of nodes, and the nodes are connected with a particular weight. These weights decide the results of the output layer (Dubrule 2003; Hampson et al. 2001). In this book, we have used as many input nodes as the number of attributes used for the analysis. Mostly, the output layer consists of one node and hence one output results, since our target is to predict one single petrophysical parameter at a time. Figure 1.8 demonstrates MLFN architecture in which four attributes are used as four-node in the input layer, one hidden layer with three nodes are used and finally one node of the output layer has been used. All nodes from the input layer are connected to the hidden layer with the weights. And all nodes of the hidden layer are connected to the output layer by the property called summation.

An alternative type of neural network is a probabilistic neural network (Masters 1995; Specht 1990, 1991). This method is actually based on a mathematical interpolation technique that uses neural network architecture for its implementation. This is the advantage of using PNN rather than MLFN since by studying the mathematical formulation one can understand the behavior in a better way.

Forgiven training datasets, the PNN technique assumes that the predicted log can be written as a linear combination of the log values in the training data. Let's we have

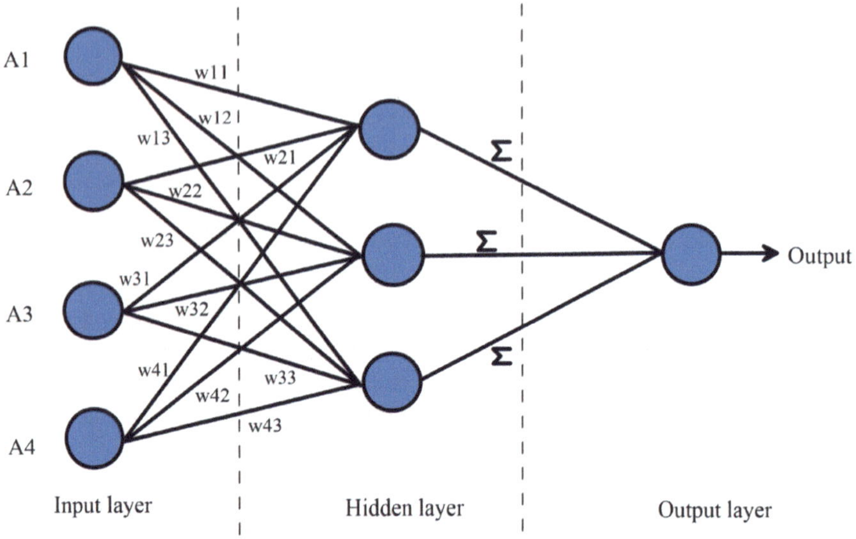

Fig. 1.8 MLFN architecture

three attributes $x = \{A_{1j}, A_{2j}, A_{3j}\}$, then the predicted log value can be estimated as follows.

$$\widehat{L}(x) = \frac{\sum_{i=1}^{n} L_i \exp(-D(x, x_i))}{\sum_{i=1}^{n} \exp(-D(x, x_i))} \tag{1.17}$$

where

$$D(x, x_i) = \sum_{j=1}^{3} \left(\frac{x_j - x_{ij}}{\sigma_j} \right)^2 \tag{1.18}$$

The quantity $D(x, x_i)$ is the distance between the input point and each of the training points x_i.

Seismic Inversion techniques have been routinely used to map the subsurface structures both in the field of seismology and seismic exploration. In seismic exploration, inversion helps extract underlying models of the physical characteristics of rocks and fluids that are helpful for reservoir characterization. Seismic inversion has several limitations. First, the seismic frequency band is limited to about 12–80 Hz, and therefore, both the low- and high frequencies in the data are missing. Well-log data provides information at these missing frequencies. Secondly, the non-uniqueness of the solution leads to multiple possible geologic models that are consistent with the observations. In addition, the seismic inversion methods ignore multiple reflections, transmission loss, geometric spreading and frequency-dependent absorption. The preferred way to reduce these uncertainties is to use additional information (mostly from well logs) containing low and high-frequency components which help to constrain the deviations of the solution from the initial-guess model. The final result, therefore, relies on the seismic data as well as on this additional information, and also on the details of the inversion methods themselves.

The details of all the discussed inversion methods are explained one by one in the coming chapters. First, the mathematical background of these inversion methods is provided and then the capacity of methods to estimate subsurface property is presented with synthetic as well as real data examples. The effort is made to explain every aspect of these seismic inversion methods and their application to the real data.

References

Aki K, Richards PG (1980) Quantitative seismology: theory and methods. New York, p 801

Ansari HR (2014) Use seismic colored inversion and power law committee machine based on imperial competitive algorithm for improving porosity prediction in a heterogeneous reservoir. J Appl Geophys 108:61–68

Bachrach R, Sayers CM, Dasgupta S, Silva J (2014) Seismic reservoir characterization for unconventional reservoirs using orthorhombic AVAZ attributes and stochastic rock physics modeling. SEG Tech Prog Exp Abs: 325–329

Banihasan N, Riahi MA, Anbaran M (2006) Recursive and sparse spike inversion methods on reflection seismic data. University of Tehran, Institute of Geophysics

Bortfeld R (1961) Approximations to the reflection and transmission coefficients of plane longitudinal and transverse waves. Geophys Prospect 9(4):485–502

Carrazzone JJ, Chang D, Lewis C, Shah PM, Wang DY (1996) Method for deriving reservoir lithology and fluid content from pre-stack inversion of seismic data. US Patent 5:583,825

Chen Q, Sidney S (1997) Seismic attribute technology for reservoir forecasting and monitoring. Lead Edge 16(5):445–448

Chunduru RK, Sen MK, Stoffa PL (1997) Hybrid optimization methods for geophysical inversion. Geophysics 62(4):1196–1207

Clochard V, Delépine N, Labat K, Ricarte P (2009) Post-stack versus pre-stack stratigraphic inversion for CO_2 monitoring purposes: a case study for the saline aquifer of the Sleipner field. In: SEG Annual Meeting, Society of Exploration Geophysicists

Connolly P (1999) Elastic impedance. Lead Edge 18:438–452

Downton JE (2005) Seismic parameter estimation from AVO inversion. In: M.Sc. thesis. University of California Department of Geology and Geophysics, pp 305–331

Dubrule O (2003) Geostatistics for seismic data integration in Earth models. Distinguished Instructor Short Course, vol 6. SEG Books

Goodway W (2001) AVO and lame constants for rock parameterization and fluid detection. CSEG Recorder 26(6):39–60

Haas A, Dubrule O (1994) Geostatistical inversion- a sequential method of stochastic reservoir modelling constrained by seismic data. First Break 12(11):561–569

Hampson DP, Schuelke JS, Quirein JA (2001) Use of multiattribute transforms to predict log properties from seismic data. Geophysics 66(1):220–236

Krebs JR, Anderson JE, Hinkley D, Neelamani R, Lee S, Baumstein A, Lacasse MD (2009) Fast full-wave-field seismic inversion using encoded sources. Geophysics 74(6): WCC177–WCC188

Lancaster S, Whitcombe D (2000) Fast-track "colored" inversion. SEG Expanded Abs 19:1572–1575

Leite EP (2010) Seismic model based inversion using matlab. Matlab-Modell Program Simul: 389

Lindseth RO (1979) Synthetic sonic logs-a process for stratigraphic interpretation. Geophysics 44(1):3–26
Masters T (1995) Advanced algorithms for neural networks: a C++ sourcebook. John Wiley & Sons, Inc.
Maurya SP, Singh KH, Singh NP (2019) Qualitative and quantitative comparison of geostatistical techniques of porosity prediction from the seismic and logging data: a case study from the Blackfoot Field, Alberta, Canada. Mar Geophys Res 40(1):51–71
Maurya SP, Singh KH (2019) Predicting porosity by multivariate regression and probabilistic neural network using model-based and coloured inversion as external attributes: a quantitative comparison. J Geol Soc India 93(2):207–212
Maurya SP, Singh NP (2018) Application of LP and ML sparse spike inversion with probabilistic neural network to classify reservoir facies distribution—a case study from the blackfoot field, Canada. J Appl Geophys Elsevier 159(2018):511–521
Maurya SP, Singh KH, Kumar A, Singh NP (2018) Reservoir characterization using post-stack seismic inversion techniques based on real coded genetic algorithm. J Geophys 39(2):95–103
Maurya SP, Singh KH (2015a) LP and ML sparse spike inversion for reservoir characterization-a case study from Blackfoot area, Alberta, Canada. In 77th EAGE Conference and Exhibition, 2015(1):1–5
Maurya SP, Singh KH (2015b) Reservoir characterization using model based inversion and probabilistic neural network. In 1st International conference on recent trend in engineering and technology, Vishakhapatnam, India
Moncayo E, Tchegliakova N, Montes L (2012) Pre-stack seismic inversion based on a genetic algorithm: a case from the Llanos Basin (Colombia) in the absence of well information. CT&F-Ciencia Tecnología y Futuro 4(5):5–20
Morozov IB, Ma J (2009) Accurate post-stack acoustic-impedance inversion by well-log calibration. Geophysics 74(5):R59–R67
Pendrel J (2006) Seismic inversion- still the best tool for reservoir characterization. CSEG Recorder 26(1):5–12
Russell B, Hampson D, Schuelke J, Quirein J (1997) Multiattribute seismic analysis. Lead Edge 16:1439–1444
Russell B, Hampson D (1991) Comparison of post-stack seismic inversion methods. In: SEG technical program expanded abstracts. Society of Exploration Geophysicists, pp 876–878
Russell B (1988) Introduction to seismic inversion methods. In: The SEG course notes, series 2
Sen R, Oltz E (2006) Genetic and epigenetic regulation of high gene assembly. Curr Opin Immunol 18(3):237–242
Sen MK, Stoffa PL (2013) Global optimization methods in geophysical inversion. Cambridge University Press, Cambridge, p 27
Sen MK, Datta Gupta A, Stoffa PL, Lake LW, Pope GA (1995) Stochastic reservoir modeling using simulated annealing and genetic algorithm. SPE Formation Eval 10(01):49–56
Shuey R (1985) A simplification of the zoeppritz equations. Geophysics 50(4):609–614
Specht DF (1990) Probabilistic neural networks. Neural Netw 3(1):109–118
Specht DF (1991) A general regression neural network. IEEE Trans Neural Netw 2(6):568–576
Vestergaard PD, Mosegaard K (1991) Inversion of post-stack seismic data using simulated annealing. Geophys Prospect 39(5):613–624
Xu S, Wang D, Chen F, Zhang Y, Lambare G (2012) Full waveform inversion for reflected seismic data. In: 74th EAGE conference and exhibition incorporating EUROPEC 2012

Chapter 2
Seismic Data Handling

Abstract Seismic data interpretation is a very crucial step and hence needs special care to the data. If some ambiguity will remain in the data, the interpretation will be false and hence loss of time and money. This chapter describes some of the processing steps of the seismic data that is necessary before proceeding towards the interpretation of seismic data. The chapter also describes the technique of picking seismic horizons, wavelet extraction and low-frequency model generation which are a very important step for the seismic inversion process.

Before moving towards the seismic inversion, seismic and well log data needs to be prepared. This preparation included seismic data conditioning, Picking of seismic horizons, wavelet extraction from seismic as well as from well log data, time to depth correlation and initial low-frequency model generation. These are very important as well as very crucial steps as the inversion results depend largely on input data. These steps are briefly described in the following sections.

2.1 Conditioning of Data

Before performing the data conditioning, if the data is in the offset domain and not in the angle domain, it needs to convert into angle gather. A full offset stack seismic gather is not modeled by the Akki-Richards equation because it is a mix of offsets, and hence cannot be used for pre-stack inversion (Singleton 2009). Further, this pre-stack seismic gathers contains significant noises that can mislead the interpretation of the inverted section. Conditioning of pre-stack seismic data is used to increase the signal to noise ration prior to pre-stack seismic inversion. Data conditioning has five major steps i.e. Band-pass Filtering, Muting, Super Gather, Probabilistic Radon Transform and Trim statics. These processes are briefly explained in the following sections along with application on real data from the Penobscot field, Nova Scotia, Canada.

S. P. Maurya et al., *Seismic Inversion Methods: A Practical Approach*, Springer Geophysics, https://doi.org/10.1007/978-3-030-45662-7_2

2.1.1 Band-Pass Filtering

The first step of data conditioning is bandpass filtering. In this process, a bandpass filter is designed and applied to a subset of seismic data. This filter suppresses unwanted low and high-frequency signals. It does not affect the wavelet shape (Mari et al. 1999). This filter is very important because sometimes the receiver record signal before reaching the main signal that is noise component and the same thing happened when the main signal is stopped and still the receiver receive some component of the signal which is again noise component and hence it needs to be separated from the main signal. Therefore a filter is designed which removes lower frequency component that is received before the main signal as well as removes the higher frequency component which is recorded by receiver after end of main signal (Yilmaz 2001). An example has been taken from the Penobscot region, Canada. Figure 2.1 shows the effect of bandpass filtering. Figure 2.1a shows before and Fig. 2.1b show after the bandpass filtering gathers. In this example, a 5/10–60/70 Hz frequency band-pass filter is applied as the gather has frequency 10–60 Hz. From the figure, it can be noticed that the resolution of the signal increases fairly and particularly, the bottom reflector which was not clear in initial seismic gather becomes very clear in bandpass filtered seismic gather.

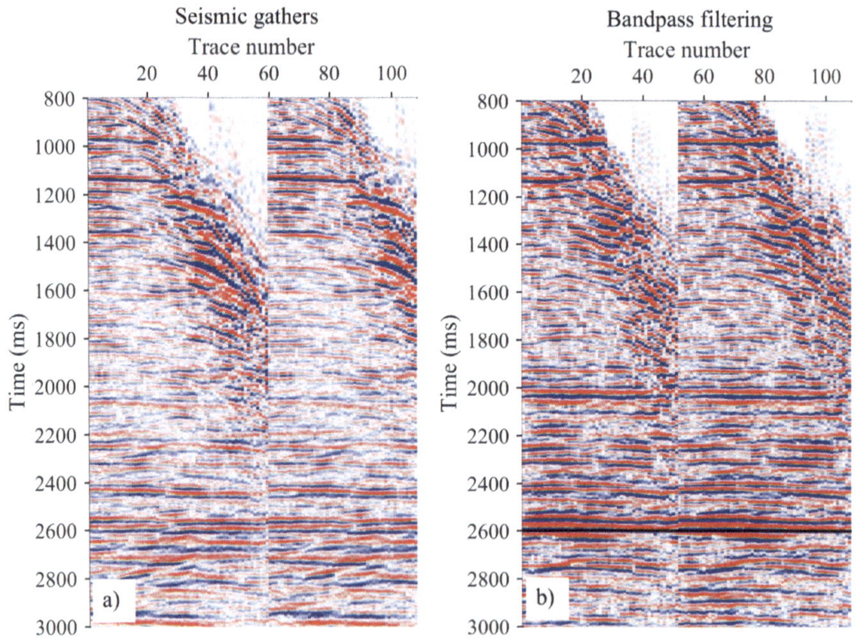

Fig. 2.1 Shows cross-section of seismic data (xline 1001 and 1002), **a** before, **b** after bandpass filtering

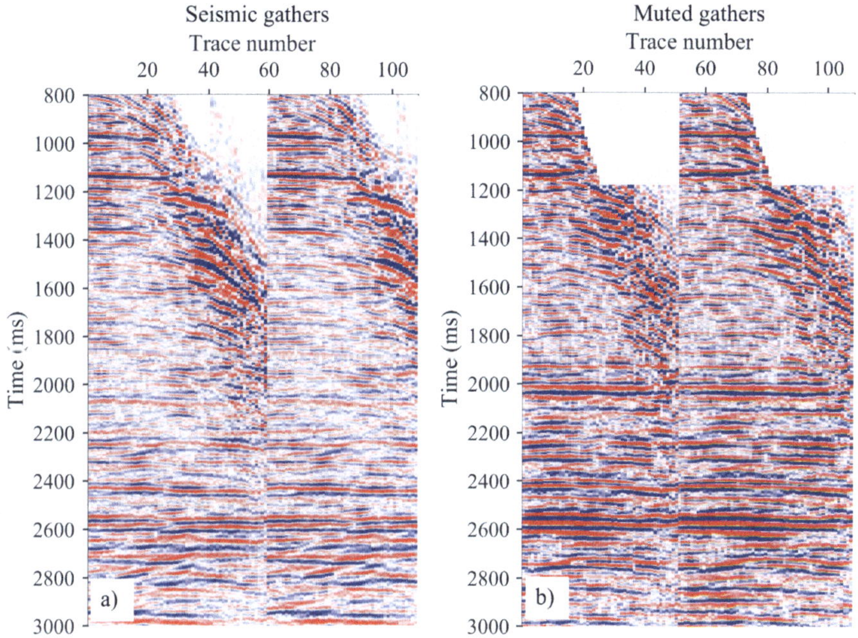

Fig. 2.2 Shows cross-section of seismic data (xline 1001 and 1002), **a** before, **b** after muting

2.1.2 Muting

Muting is the second step of data conditioning. The muting is applied to the data to eliminate noise components on the far offset traces. Muting removes bad traces by setting their amplitude to be zero. It is also applied after NMO correction to remove the effect of NMO stretch. Apart from this, the muting can also compensate for other errors, such as refractions and geometric noise (Robinson et al. 1986). Figure 2.2 shows an example of before and after muting gather which is performed on Penobscot seismic data after application of band-pass filtering. Figure 2.2a shows band-pass gather before application of the muting process and Fig. 2.2b shows seismic gather after muting. From these figures, one can notice that the far offset amplitude which is not part of the main signal is kept to be zero.

2.1.3 Super Gather

The super gather is the third step of data conditioning used to improve the signal to noise ration of pre-stack gather. The basic idea of super gather is to form average CDPs by gathering adjacent CDPs and adding them together (Mari et al. 1999; Singleton 2009). Supper gather also increases the fold of the data and eliminate

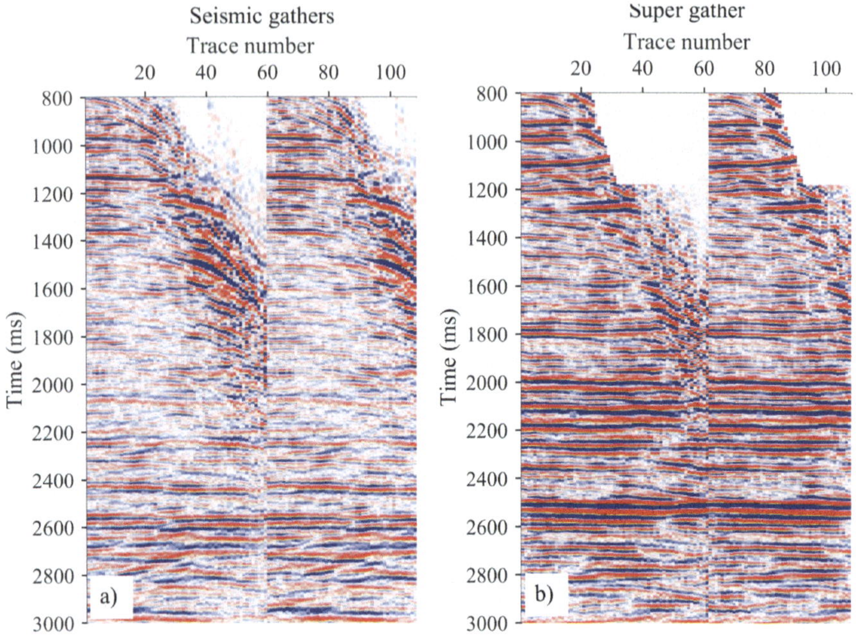

Fig. 2.3 Shows cross-section of seismic data (xline 1001 and 1002), a) before, **b** after applied super gather

noise component. This is a very crucial step of data conditioning and hence applied by expertise. Figure 2.3 describes the effect of super gather on input muted gathers. Figure 2.3a shows muted gather which is used as input and Fig. 2.3b shows supper gather results. From Fig. 2.3, one can notice that the reflector is clearer in super gather as compared to the muted gather. Sometimes seismic trace amplitude gets distorted due to external agent particularly near to reflector and hence super gather is used to correct them (Chopra et al. 2006).

2.1.4 Parabolic Radon Transform

The Parabolic Radon Filter process is the fourth step of data conditioning. The method eliminates multiple along with radon noise suppression. This is again a very important step as the pre-stack gather contains a significant amount of noise and this noisy data can mislead results and unrealistic interpretation of seismic inversion. The basic process of the radon transform is as follows. The model parameters are set to identify the long-period multiples or the random noise within the data. After the model is created, the Radon Transform then subtracts the model of these multiples or noise from the data, leaving with a data set that is greatly reduced in noise and hence optimize the traces (Yilmaz 1990; Robinson and Treitel 2000). Figure 2.4 shows an

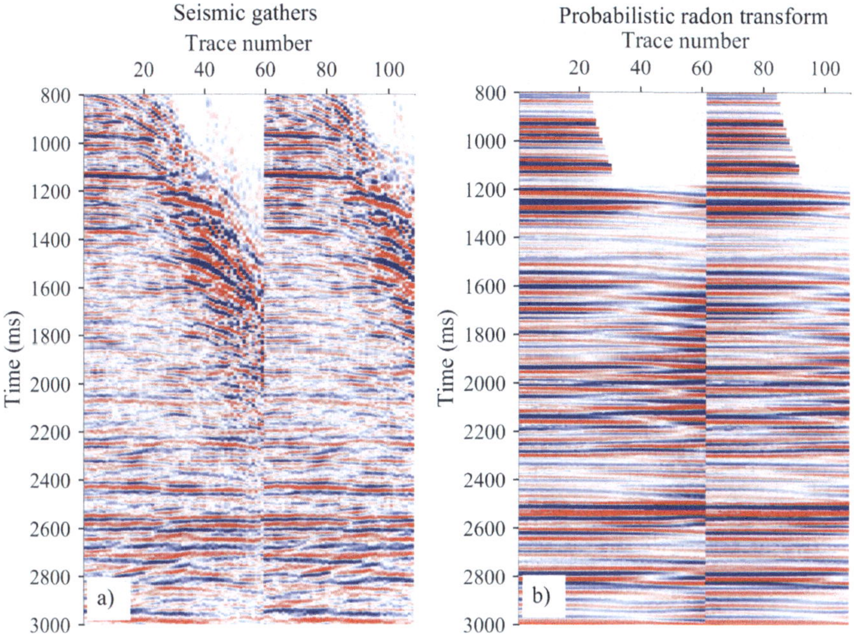

Fig. 2.4 Shows cross-section of seismic data (xline 1001 and 1002), **a** before, **b** after application of radon transform

example of a radon transform where Fig. 2.4a shows super gather before application of radon transform and Fig. 2.4b depicts pre-stack gather after radon transform. The pre-stack gather before radon transform does not reveal a consistent trend whereas after Radon filtering, the noise, apparently due to multiples, seems to be removed. It is also observed that the amount of seismic trace amplitude increase with offset. Apart from this the reflector gets more flatten in the radon transform gathers.

2.1.5 Trim Statics

The trim statics is the last step of data conditioning used to make the horizon more flatten. It also fixes migration move-out problems on pre-stack seismic gathers. The basic concept of trim statics is that it attempts to determine an optimal time shift which has been applied to each trace in the gather. The time shift is determined by cross-correlating each trace with a reference trace to make the input trace better matches that reference trace (Robinson and Treitel 2000). Usually, the reference trace is the CDP stacked trace. This step also needs the expertise to apply as in some cases it degrades the main signal and makes data look like synthetic. An example of trim statics is shown in Fig. 2.5. Figure 2.5a shows pre-stack gather before trim statics and Fig. 2.5b depicts pre-stack gather after application of trim statics. From the figure,

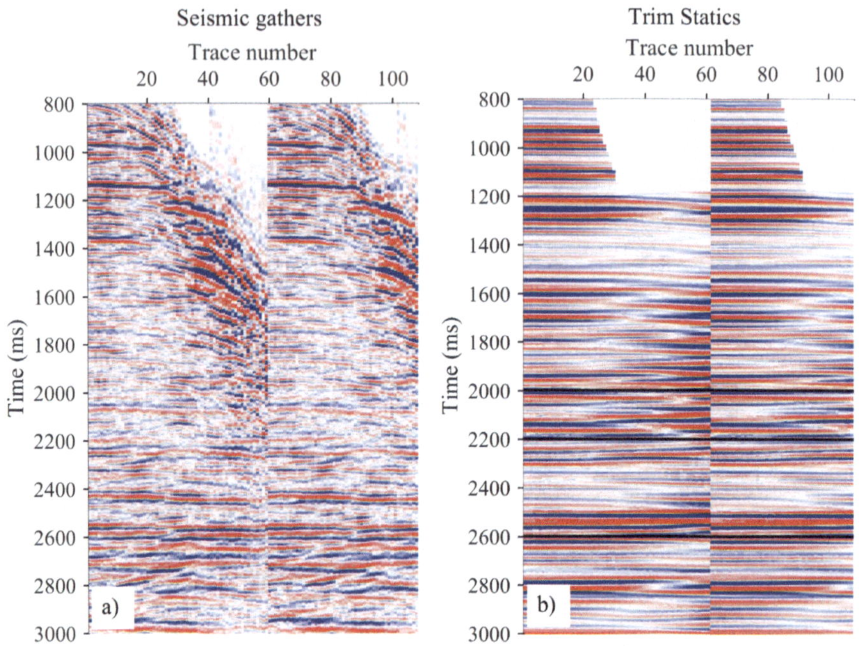

Fig. 2.5 Shows cross-section of seismic data (xline 1001 and 1002), **a** before, **b** after application of trim statics

one can see that the reflector is better aligned. The trim static was applied to the data to reduce the residual move out, up to a maximum time shift of 10 ms.

After the conditioning of gathers, the seismic data is ready for use for seismic inversion. The pre-stack seismic inversion methods are discussed in Chap. 4 using trim statics gather as input. After the preparation of seismic data, the next step involves picking seismic horizons which serve as a guide to interpolate petrophysical parameters between the wells.

2.2 Seismic Horizons

An imaginary earth's layer is usually considered as a representation of stratigraphic surfaces and the interface between two stratigraphic layers is considered as horizons. In the seismic section, these horizon shows properties of contact between two rock bodies having different seismic velocity, density, porosity, fluid content, etc. (Russell 1988). Seismic data interpretation is performed on the basis of these horizons. These horizons are very important in seismic inversion as these horizons serve as a guide to interpolate well log properties between the wells (Pendrel 2006). Picking of the seismic horizon in the seismic section is very tremendous job as they need much time and concentration. The interpretation of seismic inversion results largely depends on these horizons hence they need to perform with full attention. There are many approaches available for picking seismic horizons in 3D seismic volumes, viz. Manual picking, Interpolating, Auto picking, etc. These methods are briefly explained in the following sections. Manual picking of seismic horizons is a very old and efficient method. Manual picking is performed in interpreting horizons on inlines, cross-lines, time slices, and traverses by hand and by looking. The interpreter is looking for some degree of local continuity of the large amplitude in the data along with the local similarity of events to be picked while interpreting the horizon manually (Hildebrand and Landmark Graphics Corp. 1992).

On the other hand, the interpolation technique is somewhat more efficient than the traditional manual interpretation technique. However, the interpolation technique assumes that a horizon is locally very smooth and perhaps linear between control points. If this assumption is violated between control points then the results will be poor (Faraklioti and Petrou 2004; Pendrel 2006). This is a comparatively fast technique but in some areas like fault or joint, they need to perform with proper care.

The other horizon picking technique i.e. auto picking is nowadays very popular because the method takes the least time to pick. These methods are largely used by software developed for the interpretation of seismic data. The basic concept used in the auto picking horizon method is quite simple. The interpreter places seed picks on in-lines and/or cross-lines in the 3-D volume. These seed points are then used as initial control for the auto picking operation. If it finds such a feature within specified constraints, it picks that trace and moves on to the next trace (Dorn 1998). Simple auto pickers allow the user to specify a feature to be tracked, an allowable amplitude range, and a dip window in which to search. Graphically, how the auto picks works is explained in Fig. 2.6. The auto tracker looks for similar amplitude or similar feature in the dip window to pick and if no similar features are found then the auto tracker stops tracking at that trace (Keskes et al. 1983; Harrigan et al. 1992).

This book uses manual picking methods to pick seismic horizons in 3D seismic volume. The manual picking is the most efficient methods although the method is much time taking process. Three horizons are picked in the Blackfoot seismic data from Alberta, Canada using manual picking technique and are shown in Fig. 2.7.

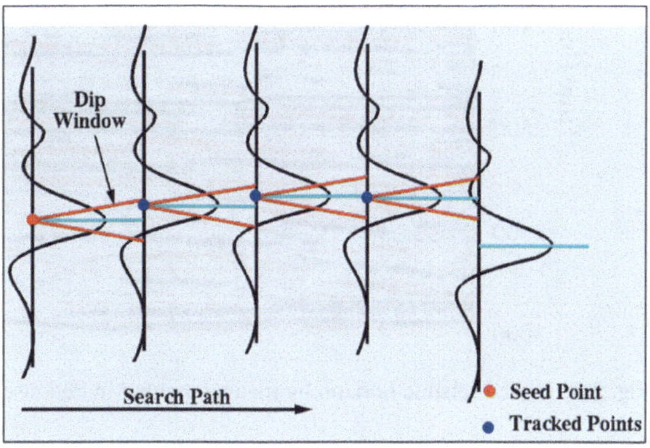

Fig. 2.6 Sketch of how an auto picker works (Courtesy SEG)

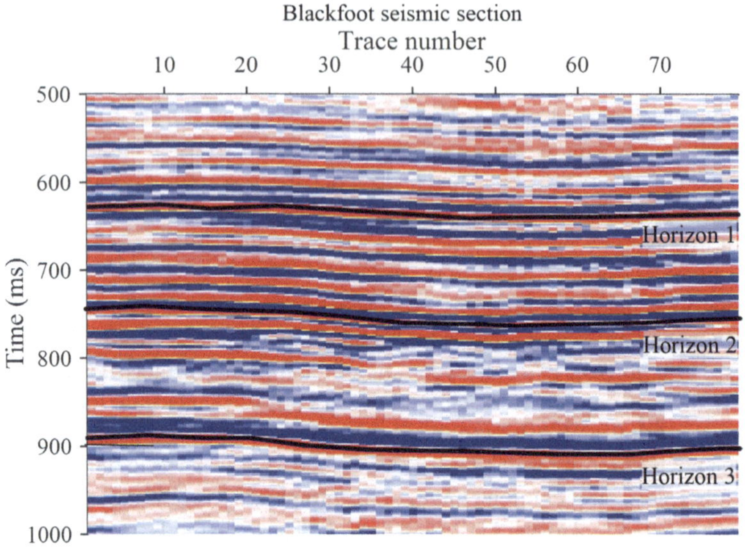

Fig. 2.7 Picked seismic horizon by manual picking in Blackfoot seismic data, Canada

Two horizons are picked in the pre-stack gather of the Penobscot region, Scotian Shelf, Canada and are shown in Fig. 2.8. It is also recommended to pick at least two horizons for getting good results from the seismic inversion.

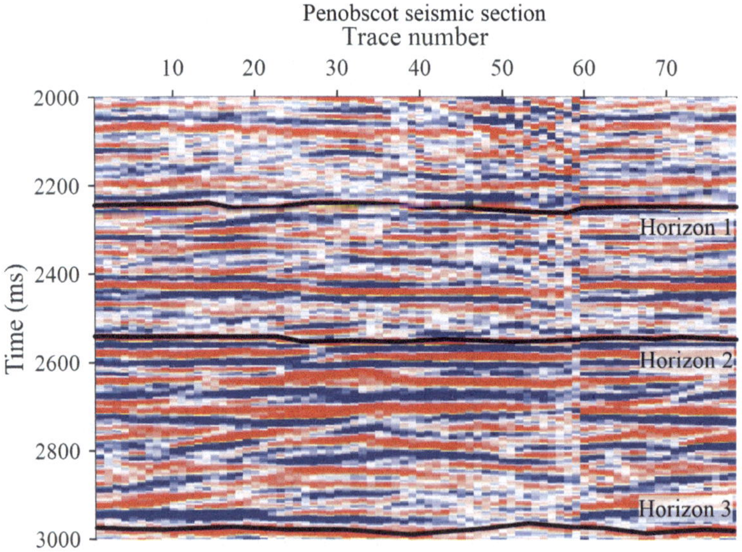

Fig. 2.8 Picked seismic horizon by manual picking in Penobscot seismic data, Canada

2.3 Seismic Wavelets

All seismic inversion methods depend on the form of forward modeling which aims to generate synthetic gather by the convolution of earth's reflectivity with source wavelet. Hence, the wavelet is a very important parameter for any type of seismic inversion. A seismic wavelet is a mathematical operator used to represent any time-varying signal into frequency varying signals. During the acquisition of seismic data, the signal sent by a source is some kind of wavelet. And if someone knows this wavelet then the seismic inversion can be performed in a more sophisticated way but unfortunately, it is unknown for most of the cases (Cheng et al. 1996). Seismic wavelet can be divided into two broad categories i.e. Zero phase wavelet and minimum phase wavelet.

2.3.1 Zero Phase Wavelet

The zero-phase wavelet has the shortest time duration and the largest peak amplitude at time zero. The wavelet is symmetrical with a maximum at time zero (non-causal). The phase of the zero-phase wavelet is zero for all frequency components contained within the signal. The fact that the energy arrives before time zero is not physically reliable but the wavelet is useful for increased resolving power and ease of picking reflection events (peak or trough) (Russell 1988; Cheng et al. 1996). A kind of zero-phase wavelet is a Ricker wavelet. Figure 2.9 shows three types of ricker wavelets, the first is the 20 Hz Ricker wavelet (Fig. 2.9a), second is 30 Hz Ricker wavelet (Fig. 2.9b) and third is 40 Hz Ricker wavelet (Fig. 2.9c) and their amplitude spectrum is shown at the bottom of wavelet. From the figure, one can be noticed that by increasing the frequency of wavelet, the wavelet gets sharper whereas the amplitude spectrum gets broader.

2.3.2 Minimum Phase Wavelet

The other wavelet which is popularly used is a minimum phase wavelet and almost all seismic sources generate minimum phase wavelet. The minimum phase wavelet has a short time duration and the maximum energy concentration at the start of the wavelet near time zero. The wavelet amplitude is zero before time zero (causal) (Russell 1988; Walden and White 1998). An example of a minimum phase wavelet is presented in Fig. 2.10. Three minimum phase wavelet is shown with frequency 10/10–40/80, 10/20–40/60 and 10/15–60/70 and corresponding amplitude spectrums are shown at the bottom of the figure. One can easily visualize the relative changes in time and amplitude spectrum due to the change of frequency. From the figure, it is noticed that all the wavelet has maximum energy concentration at near to time zero.

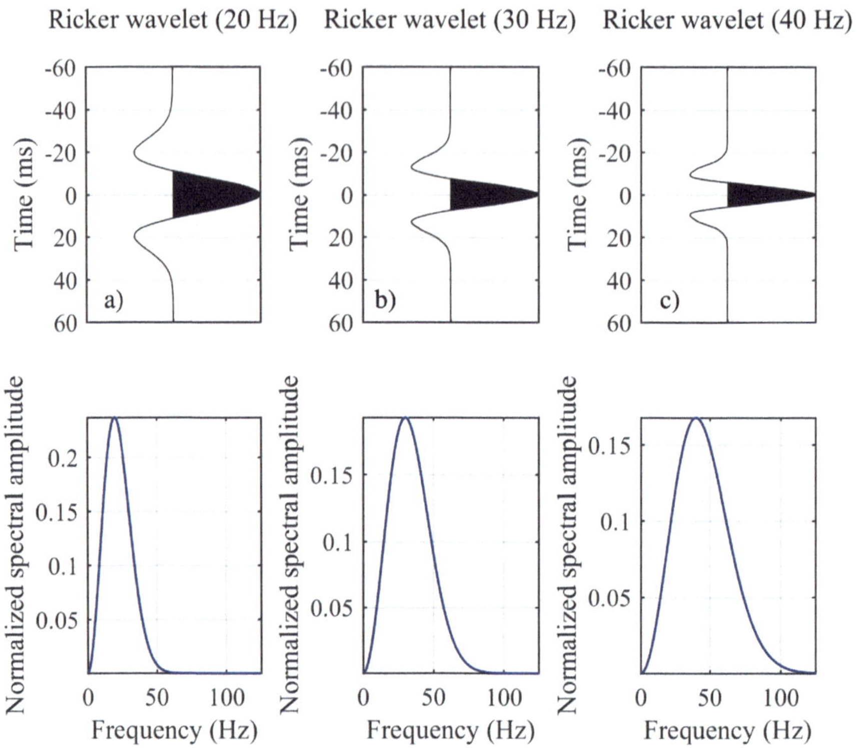

Fig. 2.9 shows zero-phase wavelet of **a** 20 Hz, **b** 30 Hz and **c** 40 Hz frequency

2.3.3 Extraction of Wavelet

Extraction of statistical wavelet from the seismic and well log data is a very important step as in most of the case; the source wavelet is unknown and almost all seismic inversion method uses wavelet which is extracted either from seismic data or well log data. Although, the wavelet extraction is a complex process, and is currently an area of active research. There are numerous wavelet extraction methods. The wavelet extraction in frequency domain consists of two parts, first determine the amplitude spectrum and second determines the phase spectrum (Swisi 2009). In these two processes, the determination of the phase spectrum is a very difficult process and includes a major source of error in seismic inversion. There are three major categories of wavelet extraction. The first is purely deterministic wavelet extraction which includes measuring the wavelet directly using surface receivers and other techniques, such as marine signatures or VSP analysis. The second approach is purely statistical wavelet extraction which includes determination of the wavelet from the seismic data alone. These procedures tend to have difficulty in determining the phase spectrum reliably. And the third approach is the extraction of wavelet using well log which uses well-log information in addition to the seismic data. In theory, this could

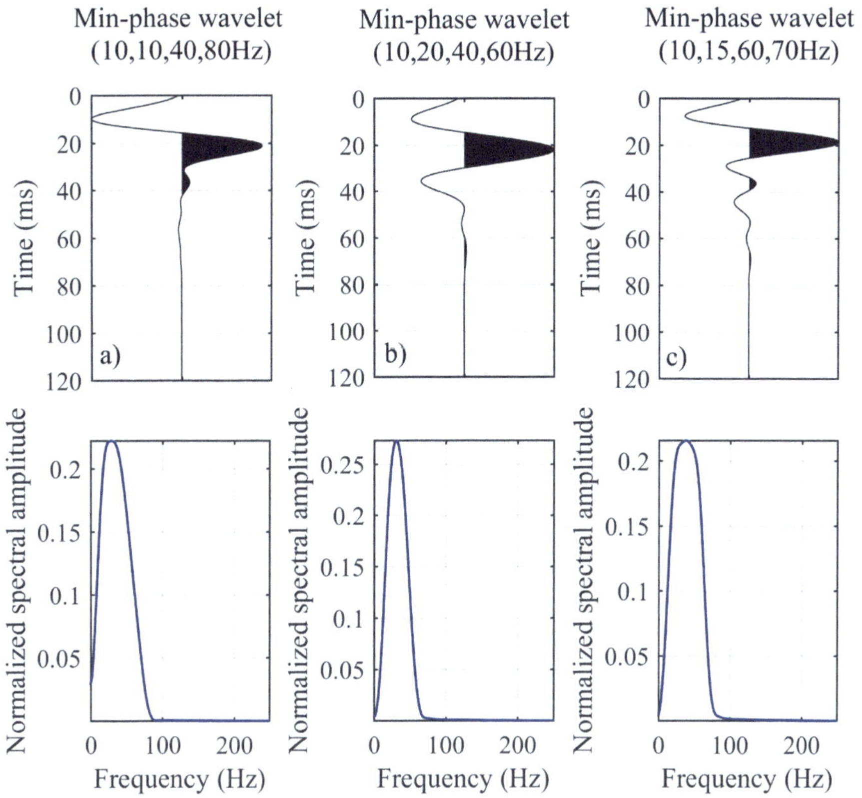

Fig. 2.10 Minimum phase wavelet of **a** 10/10–40/80 Hz, **b** 10/20–40/60 Hz and **c** 10/15–60/70 Hz frequency

provide exact phase information at the well location (Russell 1988; Swisi 2009). The problem is that this method depends critically on a good tie between the log and the seismic data. The statistical wavelet extraction and extraction of wavelet using well is most widely used and is described briefly in the following section.

i. **Statistical wavelet extraction**

The statistical wavelet extraction technique utilizes seismic data alone to extract wavelet. The mathematical procedure used here is autocorrelation. The phase spectrum of the wavelet is not calculated by this method and must be supplied as a separate parameter by the user (Edgar and Van der Baan 2011). The procedure used to extract the statistical wavelet is as following.

- Extract the analysis window in the seismic data.
- Taper the start and end of the window with a taper length.
- Calculate the autocorrelation of the data window. The length of the autocorrelation should be equal to half of the desired wavelet length.

- Calculate the amplitude spectrum of the autocorrelation.
- Take the square root of the autocorrelation spectrum. This approximates the amplitude spectrum of the wavelet.
- Add manually the desired phase.
- Take the inverse FFT (Fast Fourier Transform) to produce the wavelet.
- Sum this with other wavelets calculated from other traces in the analysis window.

ii. **Wavelet extraction using well Log**

The other type of wavelet extraction involves the use of well log data. This can be performed in two ways. The first is to use well log data to determine both; the full amplitude and phase spectrum of the wavelet and the second, it uses only well log to determine the constant phase of the wavelet in a combination of the statistical procedure described in the above section. In both approaches, the first is very popular and most widely used one (Swisi 2009). The process of wavelet extraction from the well log data is as follows.

- Extract the sonic, density, and seismic data analysis windows.
- Multiply sonic and density to get impedance and then transform it into reflectivity series.
- Apply Taper on both the seismic data and the reflectivity series at the start and end of the window.
- Next, calculate the least-squares shaping filter, W, which solves the following equation:

$$Trace(S) = W * Reflectivity(R)$$

- Thereafter, Seismic to well log correlation is performed to calculate the amplitude envelope of the wavelet using the Hilbert transform approach. If the envelope peak is not at time zero then shift the cross-correlation between log data and seismic trace and hence recalculate the wavelet using step discussed in d. This process corrects the random time shift between seismic trace and well log data.
- Take the average of the wavelet calculated by each trace to produce the desired wavelet.
- Stabilize the calculated wavelet by filtering high-frequency components.

iii. **Example of extracted wavelet**

An example of wavelet extraction is shown in Fig. 2.11 for Penobscot data, Scotian Shelf, Canada. The figure compares three wavelets, one is extracted from well log data, second is extracted from the raw seismic section and the third is extracted from trim static gathers. The corresponding amplitude spectrum is shown at the bottom of the same figure.

Further, the extracted wavelet from well log of the Blackfoot region is shown in Fig. 2.12. The area has 13 well logs hence thirteen wavelets are extracted. Wavelet

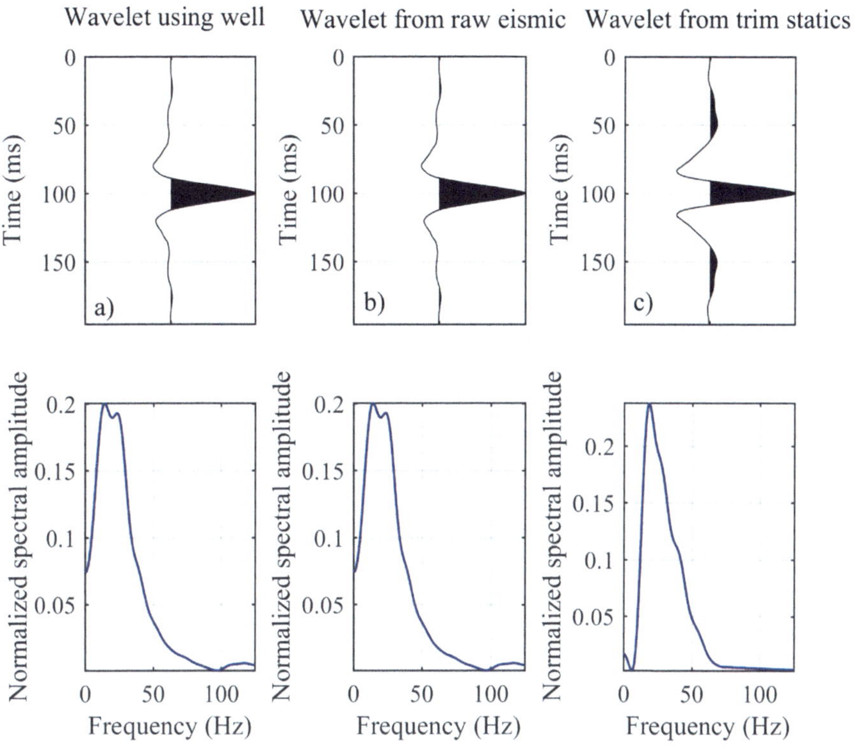

Fig. 2.11 Extracted wavelet from **a** well logs, **b** raw seismic and **c** trim gathers

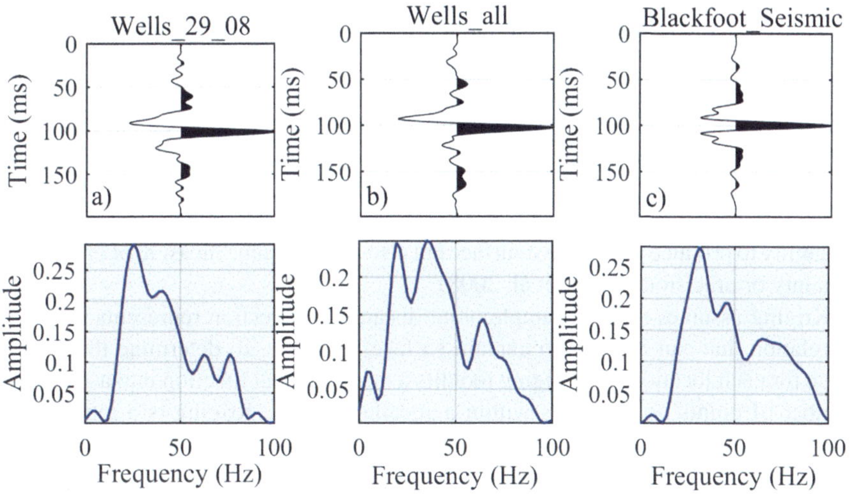

Fig. 2.12 Extracted wavelet from **a** well 29-08, **b** combine all wavelets together and **c** Blackfoot seismic

from well 29-08 is shown in Fig. 2.12a. Figure 2.12b shows the average of all wavelets extracted from all thirteen wells. Wavelet extracted from the Blackfoot post-stack seismic section is presented in Fig. 2.12c.

2.4 Low-Frequency Model

Another input used by the seismic inversion method is a low frequency model. The seismic data is band limited and contains the only frequency of ranges 10-80 Hz and hence have a lack of low and high-frequency component. These low-frequency components are very important in terms of interpretation as this low frequency contains information of deeper reflector. During the inversion of seismic data, the low-frequency component is supplied by the low-frequency model to get a broadband frequency spectrum of inverted results. The low-frequency model is obtained by the lateral interpolation and extrapolation of well log property between the wells using a seismic horizon as a guide (Swisi 2009; Maurya and Singh 2018a, b). This interpolation and extrapolation are usually performed using a variety of mathematical methods such as weighted inverse-distance, spline, kriging, and cokriging. Some of these methods are explained briefly in the following sections.

2.4.1 Kriging

Kriging is a fundamental method of statistical interpolation. The inverse distance weighted (IDW) and Spline interpolation instruments are referred to as deterministic interpolation techniques because they are based directly on the adjacent measured values or on defined mathematical formulas that determine the smoothness of the resulting surface (Todorov 2000; Chambers and Yarus 2002). A second family of interpolation techniques comprises of geostatistic techniques, such as kriging, based on statistical models including autocorrelation that is, the statistical relationships between the measured points. For this reason, geostatistical methods not only have the ability to produce a predictive surface but also provide some measure of prediction certainty or precision (Oliver et al. 2008).

Kriging assumes that the sample point distance or direction represents a spatial correlation that can be used to explain surface variation. To determine the output value for each location, the Kriging tool fits a mathematical function into a specified number of points or all points within a specified radius. Kriging is a multi-stage method involving exploratory statistical data analysis, variogram modeling, surface creation, and (optionally) exploration of a surface of variance. If you understand there is a spatially correlated distance or directional bias in the information, Kriging is most suitable (Chang et al. 1998). In soil science and geology, it is often used.

The objective is to estimate a specific property z_0^* by a linearly weighted sum of the known (measured) values at an unmeasured place. A linear estimator, written as an equation for N known values z_i, i = 1, …, N, is as follows.

$$z_0^* = \sum_{i=1}^{N} w_i z_i + w_0 \tag{2.1}$$

At each location, the estimation error is the difference between the estimated value, z_0^*, and the true value, z_0.

2.4.2 Cokriging

We often have two separate measurements in petroleum exploration: well log curves or key samples and a set of seismic information. The physical property of the subsurface, such as the acoustic and shear velocities, density, neutron porosity, etc., can be measured straight at the well places. We can interpolate and map these readings between the reservoirs using the kriging technique. However, we would like to include the seismic information in the mapping method as they can provide the exploration region with very excellent spatial coverage (Todorov 2000; Chambers et al. 2000; Ahmadi and Sedghamiz 2008). Hence Cokriging is a technique when uses both well and seismic information and predicts values away from the boreholes.

The ordinary cokriging estimator Z_0^* is written as follows.

$$Z_0^* = \sum_{i=1}^{N} w_i Z_i + \sum_{j=1}^{M} v_j X_j \tag{2.2}$$

where Z_0^* is the estimate at the grid node; Z_i is the regionalized variable at a given location, with the same units as for the regionalized variable; w_i is the undetermined weight assigned to the primary sample Z_i and varies between 0 and 100%; X_j is the secondary regionalized variable that is co-located with the primary regionalized variable Z_i, with the same units as for the secondary regionalized variable; and v_j is the undetermined weight assigned to X_j and varies between 0 and 100%.

2.4.3 Inverse Distance Weighting (IDW)

Inverse Distance Weighting is another type of interpolation technique widely used to interpolate. The technique was created on the basis of the hypothesis that points nearer to each other have more correlations and differences than those farther away. In the IDW method, the rate of correlations and similarities between neighbors is substantially assumed to be proportional to the distance between them, which can be defined as a distance-reverse function of each point from neighboring points. It should be remembered that this technique considers the concept of adjacent radius and the associated strength to the distance reverse function as serious problems. A state in which there are sufficient sample points (at least 14 points) with an appropriate dispersion at local scale concentrations will use this technique. The primary variable influencing inverse distance interpolator precision is the energy parameter value p (Burrough and McDonnell 1998). Also important to the accuracy of the outcomes are the size of the neighborhood and the number of neighbors.

$$Z_0 = \frac{\sum_{i=1}^{N} z_i . d_i^{-n}}{\sum_{i=1}^{N} d_i^{-n}} \tag{2.3}$$

where, Z_0 is the estimation value of variable z in point I, z_i is the sample value in point I, d_i is the distance of the sample point to the estimated point, N is the coefficient that determines weight based on a distance and n is the total number of predictions for each validation case.

2.4.4 Generation of Model

The low-frequency acoustic impedance (AI) model is created using the following steps.

- Calculate the AI at well locations by multiplication of well log velocity and density.
- Pick seismic horizons in 3D seismic volume to guide the interpolation between the wells. These seismic horizons also provide structural information for the model.
- Use interpolation along the seismic horizons and between the well locations to obtain the initial low AI model.
- Use the high cut filter to remove higher frequency from the model as the model contains the only low-frequency component.

A similar approach is applied to estimate a model of velocity, density, V_P/V_S etc. In this way, one can generate a low-frequency model that is used as prior information during the seismic inversion methods to supply the low-frequency component.

The examples of low-frequency models are shown in Figs. 2.13, 2.14 and 2.15. Figure 2.13 shows the low frequency AI model of the Blackfoot field, Canada. This model is created using 13 well logs of the area and two picked seismic horizons. Apart from this, a high cut filter of 10–15 Hz is used to remove higher frequency content. Figure 2.14 shows the P-impedance model for Penobscot data, Scotian Shelf, Canada. This low-frequency model is generated by using one well (L-30) and two picked seismic horizons. A high cut frequency filter of 12–15 Hz is applied to remove higher frequency from the model. Similarly, S-impedance model is generated and shown in Fig. 2.15. The Density and V_P/V_S model is shown in Figs. 2.16 and 2.17 respectively.

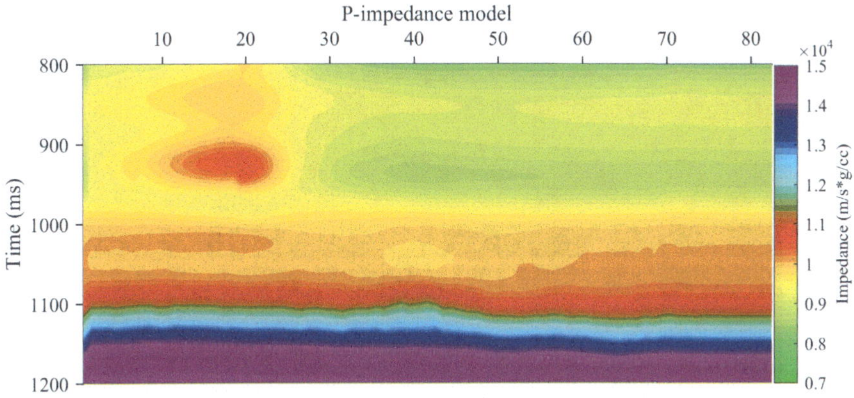

Fig. 2.13 Low-frequency AI model of the Blackfoot region, Canada

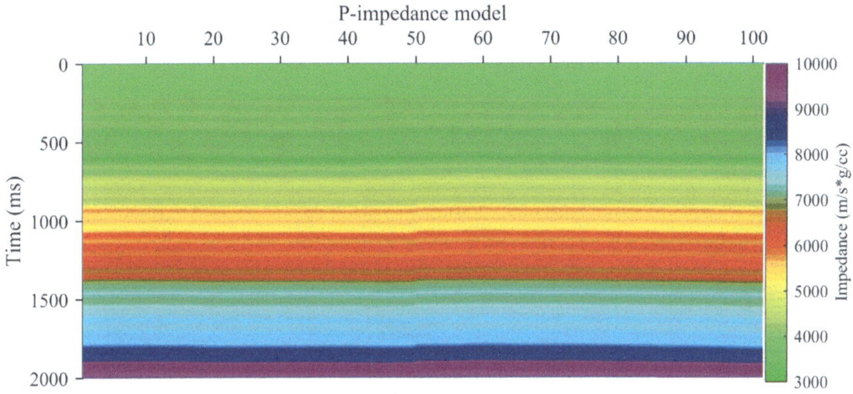

Fig. 2.14 Low-frequency AI model of Penobscot field, Nova Scotia, Canada

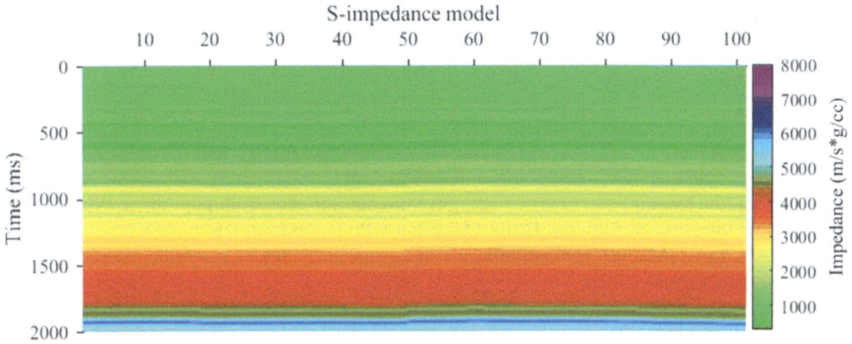

Fig. 2.15 Low-frequency S-impedance model of Penobscot field, Nova Scotia, Canada

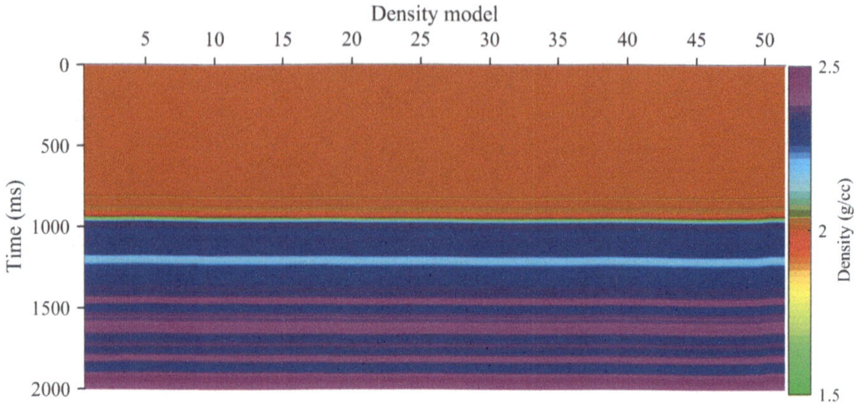

Fig. 2.16 Low-frequency density model of Penobscot field, Nova Scotia, Canada

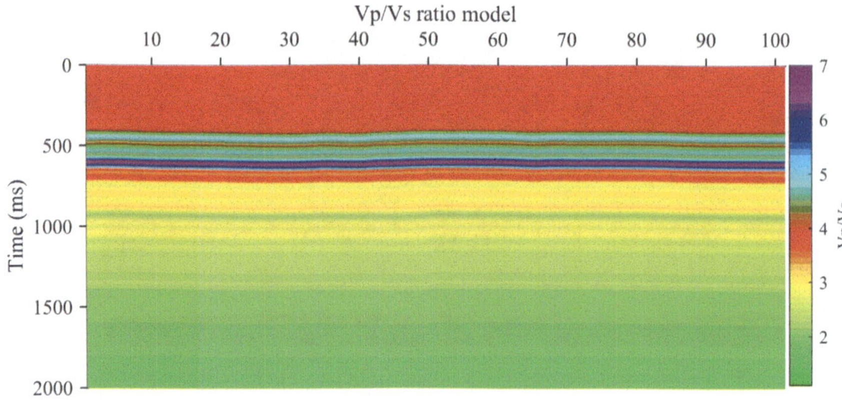

Fig. 2.17 Low frequency V_P/V_S model of Penobscot field, Nova Scotia, Canada

References

Ahmadi SH, Sedghamiz A (2008) Application and evaluation of kriging and cokriging methods on groundwater depth mapping. Environ Monit Assess 138(1–3):357–368

Burrough PA, McDonnell RA (1998) Creating continuous surfaces from point data. Principles of Geographic Information Systems. Oxford University Press, Oxford, UK

Chambers RL, Yarus JM (2002) Quantitative use of seismic attributes for reservoir characterization. CSEG Recorder 27(6):14–25

Chambers RL, Yarus JM, Hird KB (2000) Petroleum geostatistics for nongeostatisticians: part 1. Lead Edge 19(5):474–479

Chang YH, Scrimshaw MD, Emmerson RHC, Lester JN (1998) Geostatistical analysis of sampling uncertainty at the Tollesbury managed retreat site in Blackwater Estuary, Essex, UK: kriging and cokriging approach to minimise sampling density. Sci Total Environ 221(1):43–57

Cheng Q, Chen R, Li TH (1996) Simultaneous wavelet estimation and deconvolution of reflection seismic signals. IEEE Trans Geosci Remote Sens 34(2):377–384

Chopra S, Castagna J, Portniaguine O (2006) Seismic resolution and thin-bed reflectivity inversion. CSEG Recorder 31(1):19–25

Dorn GA (1998) Modern 3-D seismic interpretation. Lead Edge 17(9):1262

Edgar JA, Van der Baan M (2011) How reliable is statistical wavelet estimation? Geophysics 76(4):V59–V68

Faraklioti M, Petrou M (2004) Horizon picking in 3D seismic data volumes. Mach Vis Appl 15(4):216–219

Harrigan E, Kroh JR, Sandham WA, Durrani TS (1992) Seismic horizon picking using an artificial neural network. IEEE Int Conf Acoust Speech Signal Process 3:105–108

Hildebrand HA, Landmark Graphics Corp. (1992) Method for finding horizons in 3D seismic data. U.S. Patent, 5:153,858

Keskes N, Zaccagnino P, Rether D, Mermey P (1983) Automatic extraction of 3-d seismic horizons. In SEG technical program expanded abstracts. Society of Exploration Geophysicists, pp 557–559

Mari J-L, Glangeaud F, Coppens F, Painter D (1999) Signal processing for geologists and geophysicists. Technip, Paris, p 480

Maurya SP, Singh KH (2018a) Qualitative and quantitative comparison of geostatistical techniques of porosity prediction from the seismic and logging data: a case study from the blackfoot field, Alberta, Canada. Mar Geophys Res 40(1):51–71

Maurya SP, Singh NP (2018b) Application of LP and ML sparse spike inversion with probabilistic neural network to classify reservoir facies distribution—a case study from the blackfoot field, Canada. J Appl Geophys Elsevier 159(2018):511–521

Oliver DS, Reynolds AC, Liu N (2008) Inverse theory for petroleum reservoir characterization and history matching. Cambridge University Press

Pendrel J (2006) Seismic inversion-still the best tool for reservoir characterization. CSEG Recorder 26(1):5–12

Robinson EA, Durrani TS, Peardon LG (1986) Geophysical signal processing. Prentice-Hall International

Robinson E, Treitel S (2000) Geophysical signal analysis. Society of exploration geophysicists

Russell B (1988) Introduction to seismic inversion methods. In: The SEG course notes series 2

Singleton S (2009) The effects of seismic data conditioning on prestack simultaneous impedance inversion. Lead Edge 28(7):772–781

Swisi AA (2009) Post-and pre-stack attribute analysis and inversion of Blackfoot 3D seismic dataset. Doctoral dissertation, University of Saskatchewan

Todorov TI (2000) Integration of 3C-3D seismic data and well logs for rock property estimation. In: M.Sc. thesis, University of Calgary

Walden AT, White RE (1998) Seismic wavelet estimation: a frequency-domain solution to a geophysical noisy input-output problem. IEEE Trans Geosci Remote Sens 36(1):287–297

Yilmaz O (1990) Seismic data processing. Society of exploration geophysicists

Yilmaz O (2001) Seismic data analysis. In: Society of exploration geophysicists, vol 1. Tulsa, OK

Chapter 3
Post-stack Seismic Inversion

Abstract Post-stack seismic inversion utilizes post-stack seismic data along with well log data to estimate acoustic impedance. Post-stack seismic inversion is very fast compared to other pre-stack seismic inversion methods and provides a high-resolution subsurface image. This chapter discusses several types of post-stack seismic inversion methods namely model-based inversion, colored inversion, sparse spike inversion, and band-limited inversion. The chapter also includes the synthetic as well as real data examples of above seismic inversion methods.

3.1 Introduction

The seismic inversion techniques outlined here utilize the post-stack seismic data for the estimation of the physical properties of the subsurface. The efficacy of the inversion process relies on the effective transformation of seismic amplitudes into impedance values. Adequate precaution needs to be taken to preserve amplitude during inversion as it ensures that the measured amplitude variations are related to the geological bodies (Vestergaard and Mosegaard 1991). Thus, the multiples from the seismic data need to be processed and removed. This leads to a high signal to noise ratio and zero offset migrated section devoid of any numerical artifacts. The seismic data is band-limited, and therefore the lack of low frequencies prevents the transformed impedance traces from having the elementary velocity structure important to make geological interpretation (Clochard et al. 2009). In addition, the generated impedance volume has poor resolution incapable of identifying the thin layers. These aspects are therefore crucial and must be taken care during seismic inversion.

Post-stack techniques of inversion usually refer to the different workflows used to convert stacked seismic information into parameters of quantitative rock physics. The result is generally acoustic impedance from post-stack inversion, while pre-stack inversion can result in acoustic impedance as well as shear impedance. Sparse spike inversion, model-based inversion, recursive inversion, and colored inversion are the techniques of post-stack inversion. All of these techniques are grouped together as

S. P. Maurya et al., *Seismic Inversion Methods: A Practical Approach*, Springer Geophysics, https://doi.org/10.1007/978-3-030-45662-7_3

deterministic methods because they do not require probabilistic interventions. Post-stack seismic inversion methods used in this book are discussed in the following sections. Before the discussion of post-stack seismic inversion methods, some statistical parameters are discussed which are used to estimate the difference between the original signal and inverted signal.

3.2 Statistical Parameters

There are many statistical parameters like correlation coefficient, RMS error, synthetic relative error, peak signal to noise ratio, mean absolute error, sum absolute error, etc. to compare two signals quantitatively. These parameters can be defined as follows.

One of the most commonly used formulas in statistics is Pearson's correlation coefficient. Correlation between sets of data is a measure of how well they are related. The most common measure of correlation is the Pearson Correlation.

$$r = \frac{n\left(\sum xy\right) - \left(\sum x\right)\left(\sum y\right)}{\sqrt{\left[n\sum x^2 - \left(\sum x\right)^2\right]\left[n\sum y^2 - \left(\sum y\right)^2\right]}} \tag{3.1}$$

where n is the number of data points in the signal, x is data points of the first signal and y is data points of the second signal.

The next statistical parameter is the peak signal to noise ratio (PSNR) which can be written mathematically as follows.

$$PSNR = 20\log_{10}\left(\frac{MAX_f}{\sqrt{MSE}}\right) \tag{3.2}$$

where,

$$MSE = \frac{1}{n}\sum_{i=0}^{m-1}|x(i) - y(i)|^2 \tag{3.3}$$

and f represents the matrix data of the original signal, m and n represents number of rows and columns in the input signal.

Further, the root means square error can be defined as follows.

$$RMSE = \sqrt{\frac{\sum_{i=1}^{n}(x_i - y_i)^2}{n}} \tag{3.4}$$

The mean absolute error can be defined as follows.

$$MAE = \frac{\sum_{i=1}^{n} |x_i - y_i|}{n} \tag{3.5}$$

Absolute error is a measure of how far off a measurement is from a true value or an indication of the uncertainty in a measurement.

$$Absolute\ Error = Actual\ Value - Measured\ Value \tag{3.6}$$

One first needs to determine absolute error to calculate the relative error. The relative error expresses how large the absolute error is in comparison to the total size of the object being measured. The relative error is expressed as a fraction or multiplied by 100 and expressed as a percentage.

$$Relative\ Error = \frac{Absolute\ Error}{Known\ Value} \tag{3.7}$$

The above discussed statistical parameters are used to calculate the quantitative difference between the inverted and the original signals.

3.3 Band-limited Inversion

Band-limited inversion (BLI) is the oldest type of post-stack seismic inversion methods that utilize post-stack seismic data as input and transforms them into acoustic impedance. The acoustic impedance of the subsurface strengthens to seismic data interpretation and finding prospective zone. Further, this impedance can be transformed into many petrophysical parameters using some prediction tools like the geostatistical approach. The mathematical equation of band-limited impedance method can be derived from the relationship between the seismic trace and seismic impedance (Ferguson and Margrav 1996; Maurya and Singh 2017). The impedance is related to the velocity and density as follows.

$$Z = Velocity(v) \times density(\rho) \tag{3.8}$$

If one knows the impedance (Z), then the earth's reflectivity (r) can be estimated using the following relationship.

$$r_j = \frac{Z_{j+1} - Z_j}{Z_{j+1} + Z_j} \tag{3.9}$$

Equation 3.9 is known as normal incidence reflection formula. In Eq. 3.9, the Z_j is the seismic impedance of jth layer, and r_j is seismic reflectivity of jth and $(j + 1)$th interface. The above Eq. (3.9) can write as follows.

$$Z_{j+1}r_j + Z_j r_j - Z_{j+1} + Z_j = 0 \tag{3.10}$$

Or,

$$Z_{j+1} = Z_j\left(\frac{1+r_j}{1-r_j}\right) \tag{3.11}$$

If we put j = 2, 3, 4, …, n, one by one then we will find a series of relations as mentioned below.

$$Z_2 = Z_1\left(\frac{1+r_1}{1-r_1}\right),$$

$$Z_3 = Z_2\left(\frac{1+r_2}{1-r_2}\right) = Z_1\left(\frac{1+r_1}{1-r_1}\right)\left(\frac{1+r_2}{1-r_2}\right)$$

Similarly,

$$Z_4 = Z_3\left(\frac{1+r_3}{1-r_3}\right) = Z_1\left(\frac{1+r_1}{1-r_1}\right)\left(\frac{1+r_2}{1-r_2}\right)\left(\frac{1+r_3}{1-r_3}\right) \tag{3.12}$$

And hence one can write Eq. 3.5 for nth value as follows.

$$Z_n = Z_1\left(\frac{1+r_1}{1-r_1}\right)\left(\frac{1+r_2}{1-r_2}\right)\cdots\left(\frac{1+r_{n-1}}{1-r_{n-1}}\right) \tag{3.13}$$

From Eq. 3.13, it is noticed that the acoustic impedance for the first layer needs to be estimated from a continuous layer above the target area and then the impedance of another layer can be estimated. In this method, the impedance for the *j*th layer can thus be calculated as follows.

$$Z_{j+1} - Z_j \prod_{k=1}^{j}\left(\frac{1+r_k}{1-r_k}\right) \tag{3.14}$$

Now, the Eq. (3.14) is divided by Z_j and then taking logarithm on both sides, we get the equation.

$$\ln\left(\frac{Z_{j+1}}{Z_j}\right) \le \left(\sum_{k=1}^{j}\ln\left(\frac{1+r_k}{1-r_k}\right) \approx 2\sum_{k=1}^{j} r_k\right) \tag{3.15}$$

Thereafter, Eq. 3.15 can be solved for Z_{j+1} by considering small values of r we get the formulae.

$$Z_{j+1} = Z_1 \exp\left(2\sum_{k=1}^{j} r_k\right) \quad (3.16)$$

Further, model the seismic trace as scaled reflectivity then one can write $S_k = 2r_k/\gamma$, and hence Eq. 3.16 can be written as follows.

$$Z_{j+1} = Z_1 \exp\left(\gamma\sum_{k=1}^{j} S_k\right) \quad (3.17)$$

Equation 3.17 integrates the seismic trace and then exponentiates the result to provide an impedance trace. Band-limited impedance method uses this equation to invert the seismic trace (Waters and Waters 1987; Ferguson and Margrave 1996; Maurya and Sarkar 2016).

From Eq. 3.17, one can notice that the seismic trace gets inverted for impedance directly although seismic data have band-limited frequency and hence inverted impedance also have band-limited nature. To get broadband spectrum, a low-frequency initial impedance model is added to the inverted results. The generation of low-frequency impedance model is already discussed in Chap. 2. An important limitation of the band-limited impedance inversion is that seismic data must be in zero phases. The seismic pre-stack data could be transformed into zero phases and then can be utilized for band-limited impedance inversion (Maurya and Singh 2015b). Figure 3.1 depicts the flowchart of band-limited impedance inversion methods. The figure demonstrates that the inversion method utilizes seismic and well log data as input and acoustic impedance is estimated in the output. This method is very fast and still in use by many oil and gas companies to understand subsurface features.

3.3.1 Application of BLI to Synthetic Data

To understand more clearly the band-limited impedance inversion, an example from the synthetic data is presented. The synthetic seismogram is generated by considering 7 layers (Table 3.1) earth model and a 30 Hz Ricker wavelet. Table 3.1 illustrates the velocity, density, and depth of each layer which is used as input to generate synthetic seismogram. Figure 3.2 depicts inversion results derived using a band-limited impedance inversion approach. The track 1 shows the geological model used to generate synthetic seismograms; track 2 depicts reflectivity series estimated from the geological model; track 3 shows synthetic seismogram, and Track 4 shows the comparison of original impedance with inverted impedance along with initial guess model. From the figure, one can notice that the inverted impedance is following the trend of original impedance very nicely. The correlation is estimated to be 0.99 and RMS error is 0.067. The synthetic example case shows good performance of the algorithm as the method is able to estimate AI quite nicely.

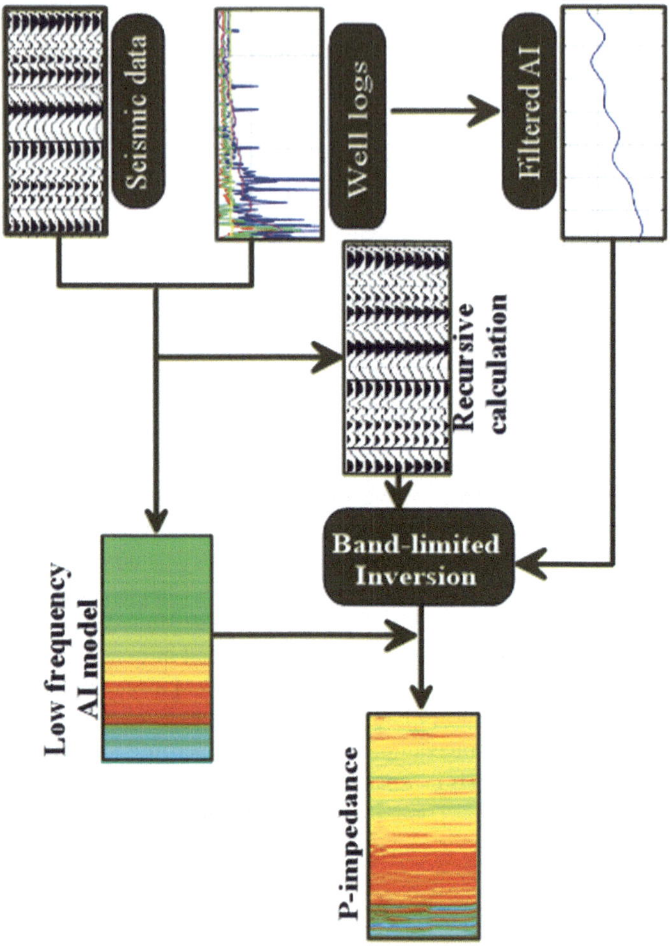

Fig. 3.1 Flowchart of band-limited impedance inversion methods

Table 3.1 Layering for synthetic modeling

Layers	Depth (m)	V_p (m/s)	V_s (m/s)	Density (g/cc)
Layer 1	0–80	3500	2000	2.3
Layer 2	80–140	4000	2300	2.5
Layer 3	140–200	3300	1900	2.2
Layer 4	200–300	3800	2200	2.4
Layer 5	300–380	4400	2500	2.6
Layer 6	380–440	3400	1960	2.3
Layer 7	440–500	4100	2200	2.5

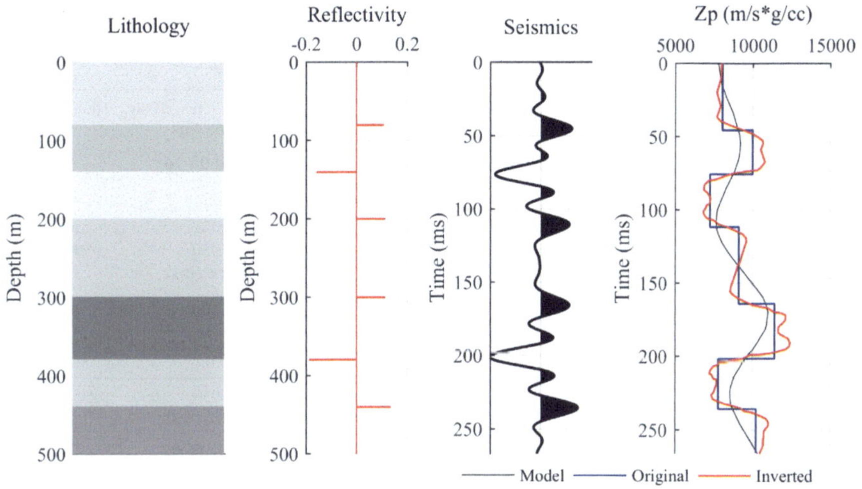

Fig. 3.2 Application of BLI methods to the synthetic seismogram

3.3.2 Application of BLI to Real Data

The other example of BLI to real field data is from the Blackfoot region, Alberta Canada which is used to estimate subsurface acoustic impedance.

Post stack seismic data along with 13 well logs are used as input for BLI. The method is applied in two steps, in the first step, composite traces near to well locations are extracted and the algorithm is applied to these traces to get AI, and in the second step, the entire seismic volume is inverted. This is standard procedure of performing inversion as these methods are time-consuming and hence one needs to optimize all the parameters of the inversion. The other benefit of using this is to cross-validate inversion results for real data case as one knows the AI from the well log data and hence inverted AI can be cross-checked by the original AI. Figure 3.3 shows comparison of inverted acoustic impedance with original AI from well log along with initial guess model for 12 wells. From Fig. 3.3, it is noticed that the inverted acoustic impedance is following well log curve very nicely for almost all wells.

It is also noticed that some well log curves have not so nicely match of phase but the overall trend is matching nicely.

Thereafter, quality checks of inversion results need to be performed and for that, the cross plot of inverted and original acoustic impedance is generated which is presented in Fig. 3.4. The figure demonstrates the cross plot of all 12 wells. The scatter points lie near to the best fit line for all the wells which indicate the good inversion results. A small cluster of scatter data lies away from the best fit line which is due to the different phases of the inverted curve in some regions. These analyses show that the method works nicely for composite trace and generates good results for the inversion of seismic volume.

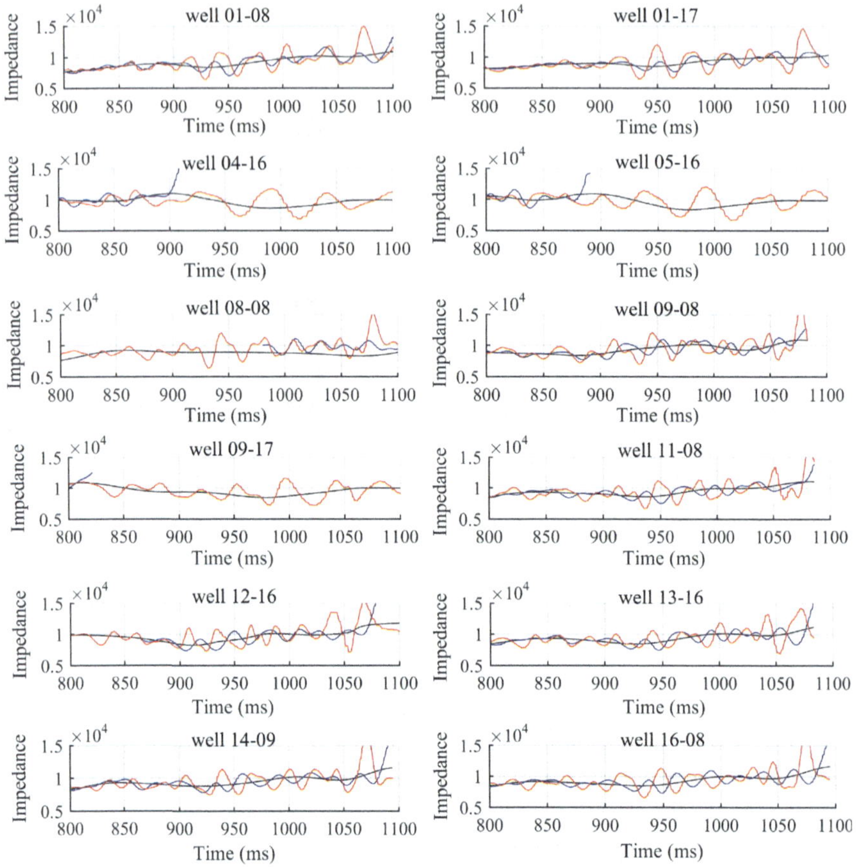

Fig. 3.3 Inversion analyses plot for 12 wells. The red curve shows inverted AI, the blue curve shows the original AI and the black curve shows the initial AI model

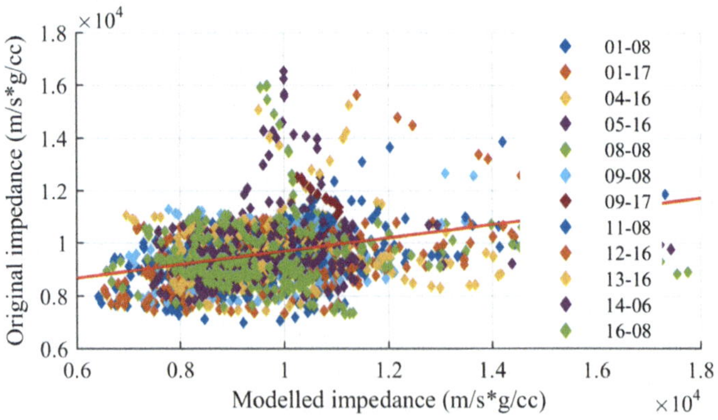

Fig. 3.4 Crossplot of modeled and original impedance for all thirteen wells

The correlation coefficient and RMS errors are estimated between inverted and original impedance from all the 13 wells and results are shown in Fig. 3.5. These values indicate the quantitative capability of BLI methods. The average correlation is estimated to be 0.89 and the average RMS error for all the thirteen wells is 1123 m/s*g/cc. These values justify the good performance of the algorithm.

Further, full volume inversion can be performed for Blackfoot 3D datasets to estimate acoustic impedance volume in the inter-well region. A cross-section of the inverted impedance (inline 28) is shown in Fig. 3.6. The variations of AI in the subsurface vertically and horizontally are clearly visible. The figure shows that the area have impedance variation from 5000 to 20,000 m/s*g/cc. The figure shows

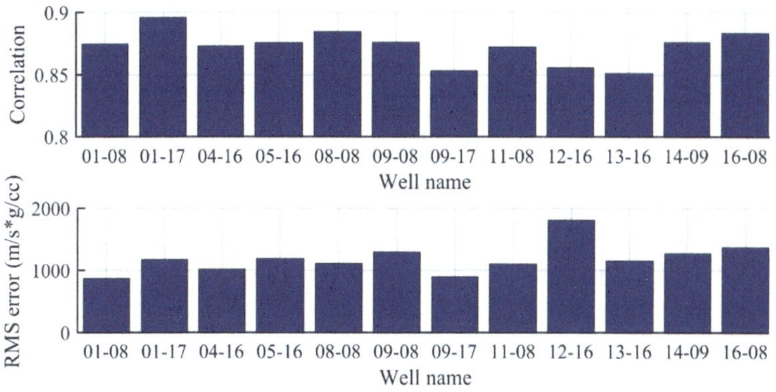

Fig. 3.5 Variation of correlation coefficient (top) and RMS error for all thirteen wells of the Blackfoot field, Canada

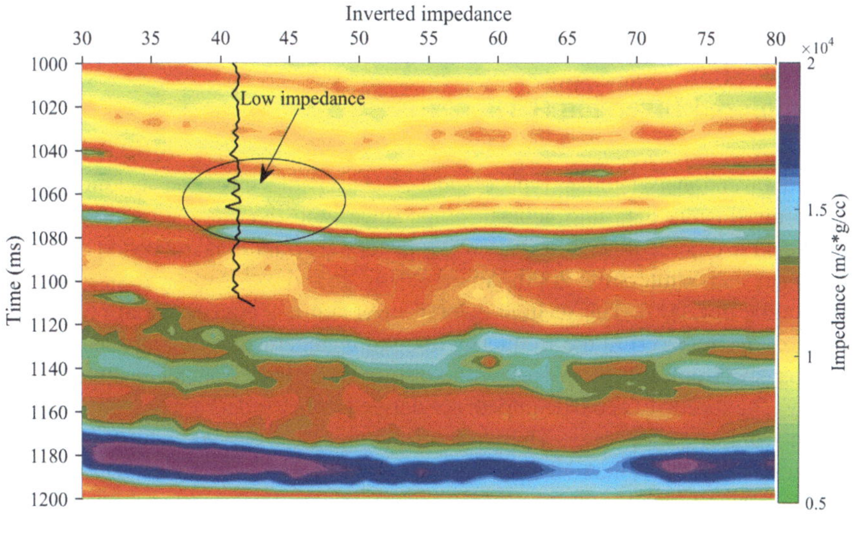

Fig. 3.6 Cross-section of acoustic impedance estimated using the BLI method

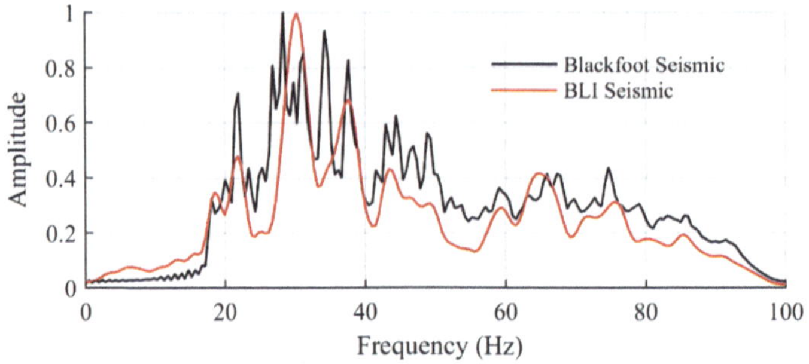

Fig. 3.7 Comparison of the amplitude spectrum of Blackfoot seismic and inverted synthetics from BLI methods

high-resolution images of the subsurface as compared to the low-resolution images of seismic data. These inverted sections enhance interpretability because of the seismic data interface property whereas inverted acoustic impedance is the layer property. The good log impedance is also plotted over the inverted section which again shows a nice match between them. The inverted section also indicates a low acoustic impedance zone in between 1060 and 1075 ms time interval. This low impedance zone may correspond to the sand channel. The anomaly zone is highlighted by the ellipse.

Sometimes, this inversion can distort the frequency content of the seismic spectrum and hence one can look at these frequencies content before interpreting these inverted sections. The comparison of the Blackfoot seismic amplitude spectrum and inverted synthetic amplitude spectrum are shown in Fig. 3.7. From Fig. 3.7, it can be noticed that the inverted sections have not all the frequency content as in the input data although, the range of frequency is the same. This can be considered as a drawback of the band-limited inversion method.

3.4 Colored Inversion (CI)

Colored inversion is another type of post-stack seismic inversion technique that transforms seismic data into acoustic impedance with the integration of well log data. Seismic colored inversion is relatively faster than traditional inversion methods (Lancaster and Whitcombe 2000; Ansari 2014). It is developed by Lancaster and Whitcombe in (2000). The seismic colored inversion is a convolutional process in which an operator (O) in the frequency domain is generated by using seismic and well log spectrum and convolves with seismic trace to get acoustic impedances directly. The spectra of acoustic impedance derived from log data are used to compute the spectrum of the operator. The phase of this operator is −90° which allows its integration with the reflectivity series to generate the impedances (Lancaster and Whitcombe 2000; Brown 2004).

Colored inversion is designed to approximately match the average spectrum of inverted seismic data with the average spectrum observed impedance (Lancaster and Whitcombe 2000). The earth's reflectivity can be considered fractal, and the resulting amplitude spectrum favors the high frequencies. If there was no preferred frequency, then one would have a white spectrum, but as there are some frequencies with more energy, then it is called colored (Swisi 2009; Maurya and Singh 2017).

The main process of colored inversion is to design colored operators. The operator is derived by the following steps.

(a) Calculate acoustic impedance and plot against frequency for all the wells available in the area, and then a regression line is fitted to the amplitude spectrum of the acoustic impedance which represents the impedance spectrum in the subsurface on a log-log scale (Fig. 3.8a).
(b) Calculate the seismic spectrum near the well locations and select the mean spectrum (Fig. 3.8b). Thereafter, the well log spectrum in (a) and seismic mean spectrum in (b) is used to calculate the operator spectrum.

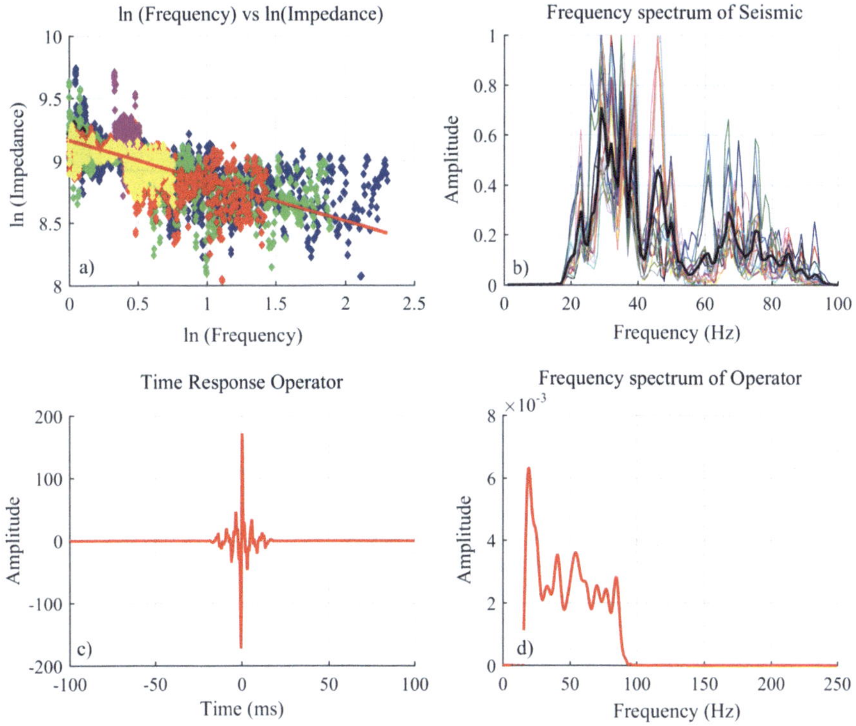

Fig. 3.8 Process of estimation of colored operator, **a** using a set of wells from the area, the amplitude spectra of the acoustic impedance for all the wells are plotted on log-log scale (I), **b** using a set of seismic traces from around the wells, the average seismic spectrum is calculated (S), **c** operator in time domain (I/S), and **d** operator in frequency domain

(c) The final spectrum needs to combine −90° phase shift to create the desired operator in the time domain (Fig. 3.8c).
(d) The operator in the frequency domain is used to transform data which can be achieved from Fourier transform (Fig. 3.8d).

Colored inversion is fast and suitable for application to 3-D datasets (Ansari 2014). Generally, traditional inversion methods are time-consuming, expensive, require specialists and are not performed routinely by the interpretation Geophysicist, whereas CI is rapid, easy to use, inexpensive, robust and does not require expert users (Maurya and Sarkar 2016). Figure 3.9 depicts flowchart of colored inversion methods. The figure shows that the colored inversion utilizes seismic and well log data and generates an operator. Further, this operator convolved to the seismic trace to get acoustic impedance as output. The method is very simple and very fast. But

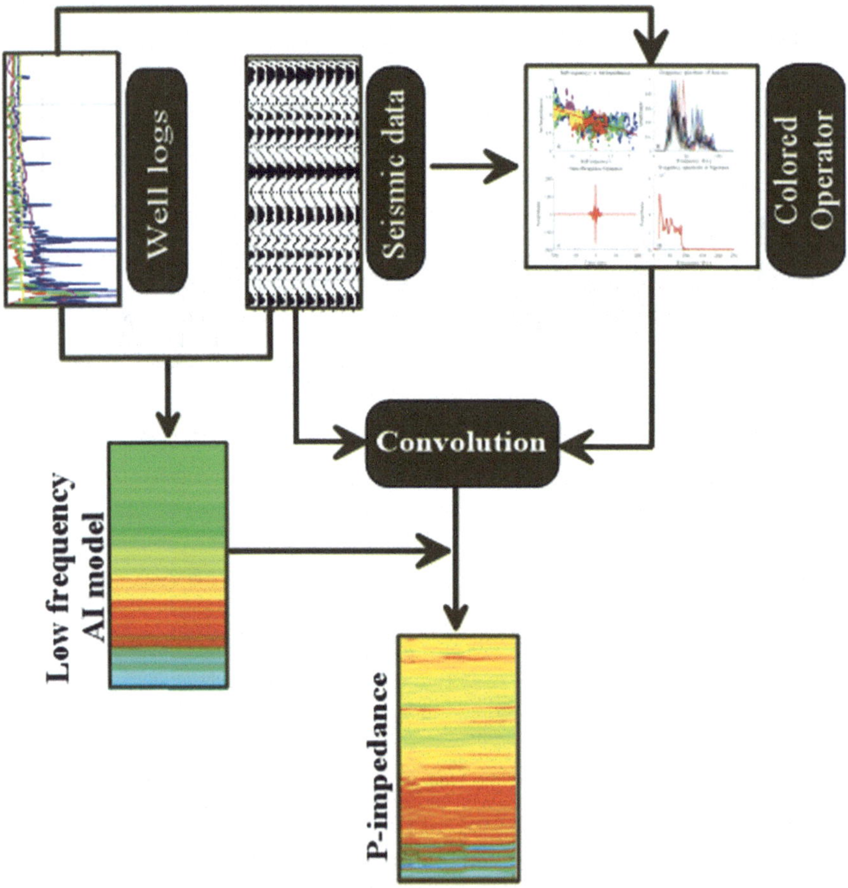

Fig. 3.9 Flowchart of colored inversion method

some author argues (Russell 1988; Maurya and Singh 2018, 2019; Maurya et al. 2019) that the colored inversion gives an average variation of impedance and hence one cannot use for the quantitative interpretation.

3.4.1 Application of CI to Real Data

Real data example from the Blackfoot data, Alberta, Canada is presented here. 3D post-stack seismic data along with 13 well logs are used as input for colored inversion methods. The method is first tested on the composite trace and then applied to the 3D seismic volume. The area has thirteen wells and hence thirteen seismic trace is extracted near to well location and colored inversion is applied one by one to all traces. Figure 3.10 shows a comparison of inverted AI with original AI from the well logs along with the initial AI guess model. The area has 13 wells but the results are shown for 12 wells for simplicity. From the figure, one can notice that all the inverted curves are matching with the well log curves very nicely. This indicates the good performance of the algorithm.

For quality check of the inversion results, a cross plot of inverted impedance with well log impedance is generated for all the wells and is shown in Fig. 3.11. Further, a best fit straight line is fitted to the data and the variation of scatter points away from the best fit line is noticed. The figure indicates that most of the scatter points lie near to the best fit line which presents that the inverted data points are very close to the original data points and hence indicate good performance of the CI.

For the quantitative comparison of inversion results for composite trace, some numerical parameters are extracted and shown in Fig. 3.13. The other parameters are given in Table 3.2. Figure 3.12 shows two histogram plots, one corresponds to the correlation coefficient (Top) and the other corresponds to the RMS error (Bottom). From the figure one can notices that the correlation coefficient varies from 0.86 to 0.89 with average value of 0.875 and RMS error varies from 700 to 1400 m/s*g/cc with average value of 1200 m/s*g/cc.

Further, 3D inversion is performed and acoustic impedance in the inter-well region is estimated. The parameters used for the colored inversion analysis are as follows.

- Colored inversion was initialized by 13 data pairs.
- Processing sample rate = 2 ms
- Threshold for inverse = 20
- Operator Length = 200
- Taper Length = 50
- Regression Line: Intercept = 5.92741, Gradient = −0.733055.

The cross-section of inverted acoustic impedance at inline 27 is shown in Fig. 3.13. The section shows variations of AI horizontally as well as vertically in the subsurface and depicts very high resolution as compared with the seismic section. The area have impedance variation from 5000 to 17,000 m/s*g/cc. The main benefit of

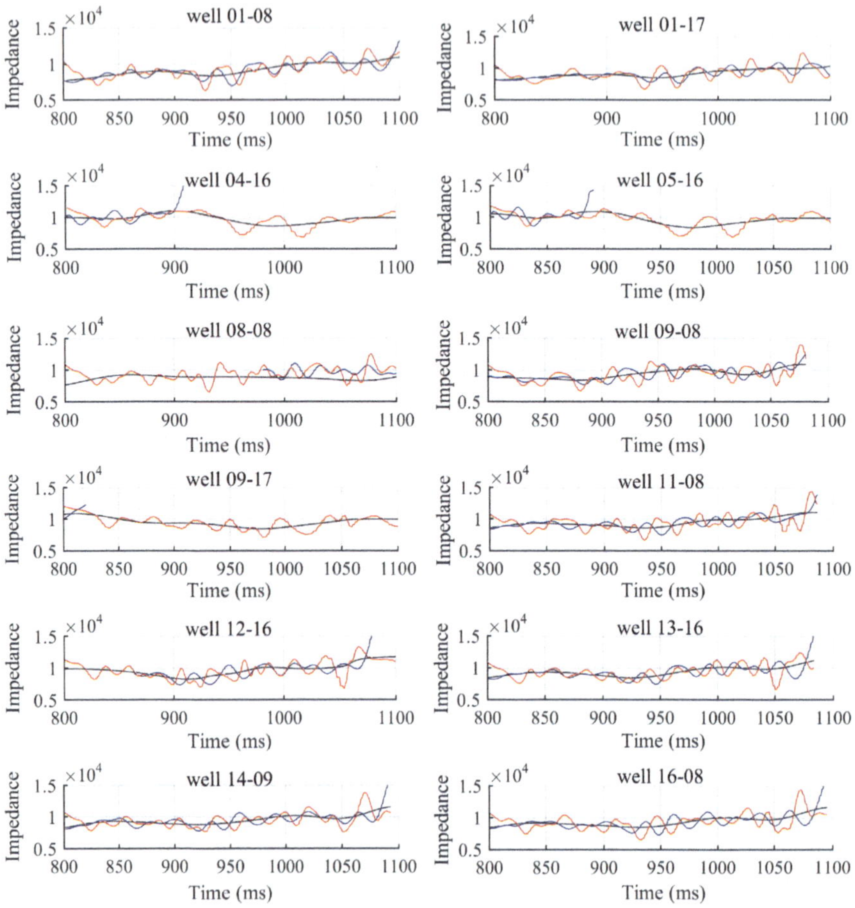

Fig. 3.10 Inversion analysis for 12 wells estimated using colored inversion methods

interpretation from the inverted AI section is that the information is given as layered whereas the seismic section gives information at the interface only. The well log impedance is also plotted above the inverted AI section for cross verification of results. Figure 3.13 indicates good matching between the well log and inverted AI section. From the inverted section, one can notice that the impedance contrast increases with increasing depth which is obvious as velocity and density increase with increasing depth. The low impedance anomaly is also clearly visible in between 1060 and 1075 ms time intervals. One can also notice that the interpretation is easier from the inverted section as compared with the seismic sections. This is the reason why exploration companies use more frequently these inversion techniques. However, one seeks to find a prospective zone then the colored inversion method is the best method as the method is less time consuming and requires less expertise.

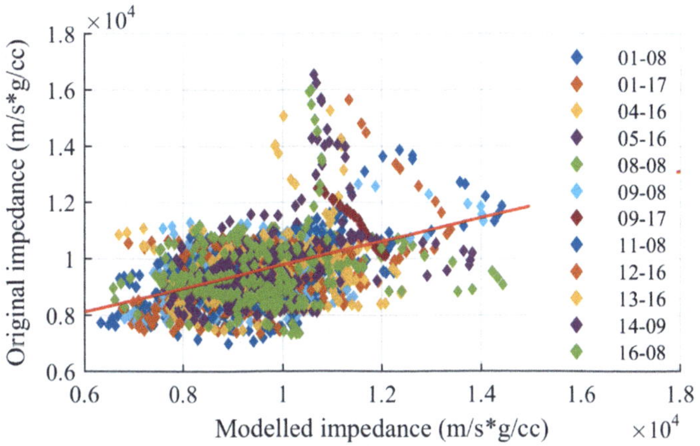

Fig. 3.11 Crossplot of inverted impedance with original impedance

Table 3.2 Quantitative comparison of seismic inversion results (AI stands for acoustic impedance, CC stands for colored inversion, PSNR stands for peak signal to noise ratio, MAE stands for mean absolute error and SAE stands for sum absolute error)

Colored inversion						
Properties	Reservoir zone	CC	RMS error	PSNR	MAE	SAE
AI	6.5×10^3–8.0×10^3	0.84	1.5×10^3	15.2	1×10^3	8×10^4
Amp. spectrum	NA	0.93	0.1	70.4	0.0	10.0

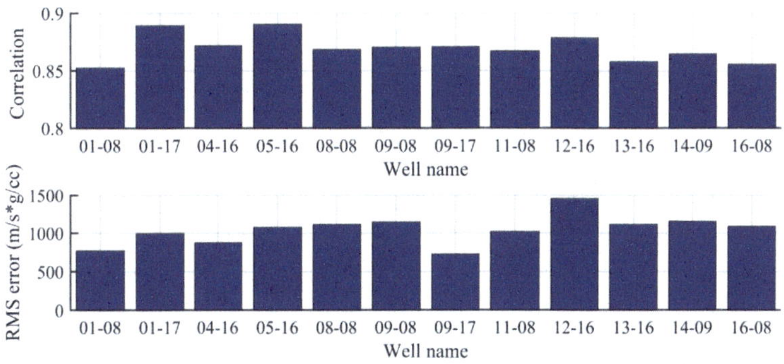

Fig. 3.12 Variation of correlation (top) and RMS error (bottom) for inverted results estimated using colored inversion methods

Further, to check frequency content in the inverted seismic, one can compare the amplitude spectrum of input seismic i.e. Blackfoot seismic and inverted synthetic seismic. The comparison is presented in Fig. 3.14. From the figure, one can notice that the amplitude spectrum of inverted synthetic seismic is not matching with the

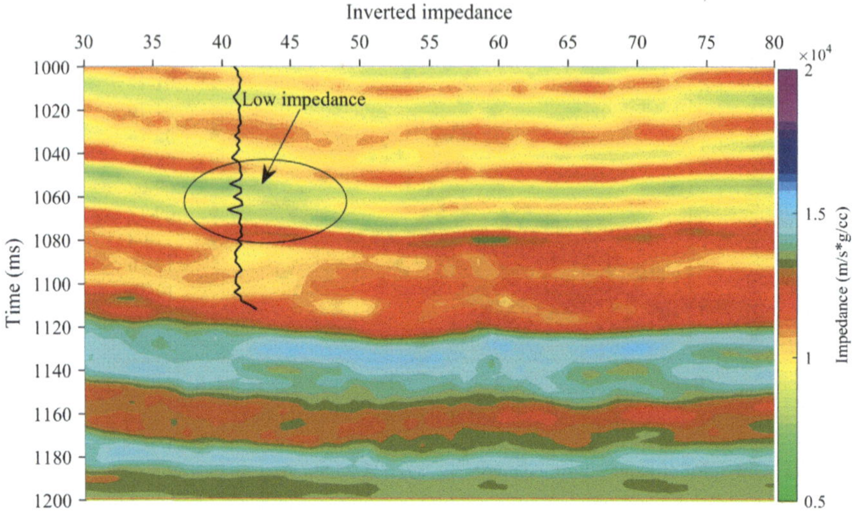

Fig. 3.13 Cross-section of inverted acoustic impedance estimated using colored inversion methods

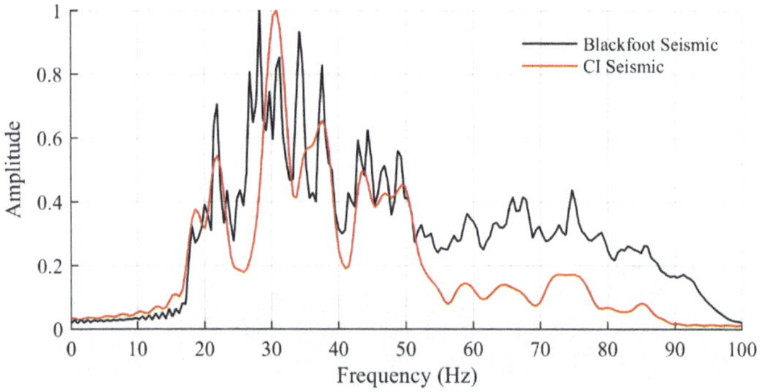

Fig. 3.14 Comparison of amplitude spectrum between Blackfoot seismic and inverted synthetic data

input seismic trace although the range of both spectra is the same. This comparison suggests that the colored inversion method distort frequency component completely during the process and hence dependency of the interpretation on inverted results gets restricted to qualitative only. The colored inversion method distorts frequency more as compared to the band-limited inversion methods. This is the drawback of seismic colored inversion methods.

In the above section, we have presented a straightforward description and application of colored inversion (CI). The method is called a robust process in the literature, but it is somewhat sensitive to the chosen frequency range. Notwithstanding all this, the process of colored inversion is simple and fast and yields informative subsurface images for the interpreters. Now, we will look at the other popular post-stack seismic inversion methods i.e. Model-based inversion.

3.5 Model-based Inversion (MBI)

The model-based inversion is a very famous post-stack inversion method in the world of geoscience. The basic concept of model-based inversion is to assume a low-frequency model of the P- impedance and then perturb this model until one obtains a good match between the seismic and computed synthetic seismic trace. It is assumed that the extracted wavelet is a good estimate of the seismic section. Seismic wavelet is a source signature and is required in most of the inversion techniques (Leite 2010; Maurya and Singh 2015). The model-based inversion technique is principally based on the convolutional theory which states that the seismic trace can be generated from the convolution of wavelet with the Earth's reflectivity series and addition of noise (Russell 1988; Mallick 1995).

$$S(t) = W(t) * r(t) + n(t) \tag{3.18}$$

where, $S(t)$ is seismic trace, $W(t)$ is wavelet, $r(t)$ is earth's reflectivity, $n(t)$ is noise component and some time it is assumed to be zero for simplicity and $*$ indicates the convolution process. If the noise in the data is uncorrelated with the seismic signal, the trace can be solved for the earth's reflectivity function (Stull 1973; Russell 1988).

The main benefit of model-based inversion is that the methods involve the updating of the geological model rather direct inversion of seismic data. This is beneficial because the impurities in the seismic data are not incorporating in the inversion results whereas the other inversion method transforms the main signal which has transmission losses, spherical divergence, unwanted signals, etc. (Ferguson 1996).

The method can be classified into two parts by questioning. The first question is; what is the fundamental mathematical relationship between the model and the seismic data and the second question is; how to update the guessing model. To answer this question, we have considered two different approaches namely, the Generalized Linear Inversion (GLI) and the Seismic Lithologic (SLIM) method. These methods are briefly discussed in the following sections.

3.5.1 Generalized Linear Inversion (GLI) Method

The model-based inversion methods use the principle of generalized linear inversion method. The basic idea behind this method is, given a set of geophysical observations, the GLI method will find the geological subsurface model (AI in this case) which best fits the observations in the least square sense. The generalized linear inversion method is developed by Coke and Schneider (1983) and it is now very common in use. This can be expressed mathematically as follows.

If one can write observations in the form of vector then the parameter of the k model can be written in vector form as follows.

$$M = (m_1, m_2, \ldots, m_k)^T \tag{3.19}$$

and observation with n data points can be written in vector form as follows.

$$T = (t_1, t_2, \ldots, t_n)^T \tag{3.20}$$

Now, one can express the relationship between the model parameters and observations in the functional form as follows.

$$t_i = F(m_1, m_2, \ldots, m_k), \text{where} i = 1, \ldots, n \tag{3.21}$$

If one is able to derive the functional relationship between the model parameters and observations, any set of model parameters will be produced as output. The GLI eliminates the need for trial and error by analyzing the error between the model parameters and the observations and then perturbing the model parameters in such a way as to produce an output which will produce less error (Yilmaz 2001). In this way, one may iterate towards a solution. Mathematically, the observation can be written by Taylor's series expansion of the forward model.

$$F(M) = F(M_0) + \frac{\partial F(M_0)}{\partial M_0}\Delta M + \frac{\partial^2 F(M_0)}{\partial M_0^2}\Delta M^2 + \cdots \tag{3.22}$$

where M_0 is initial guess model, M is true earth's subsurface model, ΔM is changed in model parameters, $F(M)$ is termed as observations, $F(M_0)$ is calculated function values from the initial guess model and $\partial F(M_0)/\partial M$ is called a change in the calculated value or sensitivity matrix (Russell 1988). The linearized version of Eq. 3.22 can be written as follows.

$$F(M) = F(M_0) + \frac{\partial F(M_0)}{\partial M_0}\Delta M \tag{3.23}$$

In Eq. 3.23, the $F(M) - F(M_0)$ is termed as an error vector generated by subtracting the synthetic seismic trace from the observed seismic trace of the area.

Again, the error vector between the observations and the computed values can be expressed as follows.

$$\Delta F = F(M) - F(M_0) \tag{3.24}$$

Equation 3.24 can be rewritten in the matrix form as follows.

$$\Delta F = G\Delta M \tag{3.25}$$

In Eq. 3.25, the G is matrix derivatives which have n number of rows and k number of columns. The tendency of this error is decreasing in a roughly exponential manner with iteration, but the condition is that the initial guess model lies within the range of convergence. The iteration is continued until the error drops below some predetermined level. One can write solution of Eq. 3.25 as follows.

$$\Delta M = G^{-1}\Delta F \tag{3.26}$$

Equation 3.26 is an overdetermined case since there is usually a number of observations than a number of parameters and hence the matrix G is usually not square matrix and therefore it is not possible to estimate true inverse of G matrix (Quijada 2009). To overcome this problem, one can use the least square solution which is also known as the Marquart-Levenburg method. With this method, one can write the solution of Eq. 3.25 as follows.

$$\Delta M = \left(G^T G\right)^{-1} G^T \Delta F \tag{3.27}$$

However, one must derive the functional relationship between model parameters and observations which is necessary to relate to each other. The simplest solution is the convolution model which can be expressed as follows.

$$S(t) = W(t) * r(t) \tag{3.28}$$

Equation 3.28 is not directly used by Coke and Schneider (1983) as this equation not include multiples, transmission losses, etc. in its implementation. Coke and Schneider (1983) use a modified version of Eq. 3.28. The main advantage of incorporating multiples and transmission loss in the solution is that they are not included in the model parameters, although they are modeled in computing the seismic response (Russell 1988). This is big advantage of GLI and hence model-based inversion method over recursive methods discussed above since those methods incorporate the multiples and transmission loss into the solution if the data contains it. Figure 3.16 displays flowchart of the GLI methods.

The workflow of model-based inversion technique is as follows (Ferguson and Margrave 1996):

1. Calculate the acoustic impedance at well locations using the well log data and Pick horizons in the seismic section to control the interpolation and to provide structural information for the model between the wells in the area (Bosch et al. 2010).
2. Use the kriging interpolation technique along with the picked seismic horizons to obtain the initial acoustic impedance model (Maurya and Singh 2015).

3. Extract statistical wavelet from the seismic section and convolve it with the Earth reflectivity (generated from the initial impedance model) to obtain synthetic seismic trace. This synthetic trace is different from the observed seismic trace.
4. The least-squares optimization is performed for minimizing the difference between the real and modeled reflectivity section. This is achieved by analyzing the misfit between the synthetic trace and the real trace and modifying the block size and the amplitude to reduce the error (Brossier et al. 2015; Maurya and Singh 2017).

3.5.2 Seismic Lithologic Modelling (SLIM)

The Seismic Lithologic methods are described in Cooke and Schneider (1983) but the developed methods are not commercially used as it is not promising for getting always great results. However, Western Geophysical developed a similar technique of Seismic Lithologic Modeling (SLIM) which is commercially used as part of model-based inversion. Although the Western Geophysical's developed algorithm has not been fully published and hence the mathematics behind it is not in the public domain. The SLIM technique involves a model being perturbed rather than a seismic segment being directly reversed.

The flowchart of SLIM is presented in Fig. 3.15. The basic idea behind the technique is that the original geological model is produced and compared to a seismic section, as in the GLI technique. The model is described at different control points along the line as a sequence of layers of variable velocity, density, and thickness. Also, either the seismic wavelet is provided or it is estimated. The synthetic model is then compared to the seismic data and the amount of the least squared error is calculated. The model is perturbed in such a manner that the error is reduced, and until the convergence, the method is continued.

The user has complete power over the limitations and can integrate any source of geological data. The main benefit of this technique over conventional recursive techniques is that noise is not integrated into the process.

3.5.3 Application of MBI to Synthetically Generated Data

Here two examples are presented to examine the capability of MBI to estimate acoustic impedance, first is from synthetic data and second is real data from the Blackfoot field, Canada. To generate a synthetic seismogram, we have utilized velocity and density as a geological model given in Table 3.1. And then, synthetic seismograms are generated over the 7 layer earth model by convolving 30 Hz ricker wavelet to the reflectivity. Figure 3.16 shows inversion results for 7 layer earth model. Track 1 of Fig. 3.16 shows the assumed geological model, track 2 shows calculated reflectivity

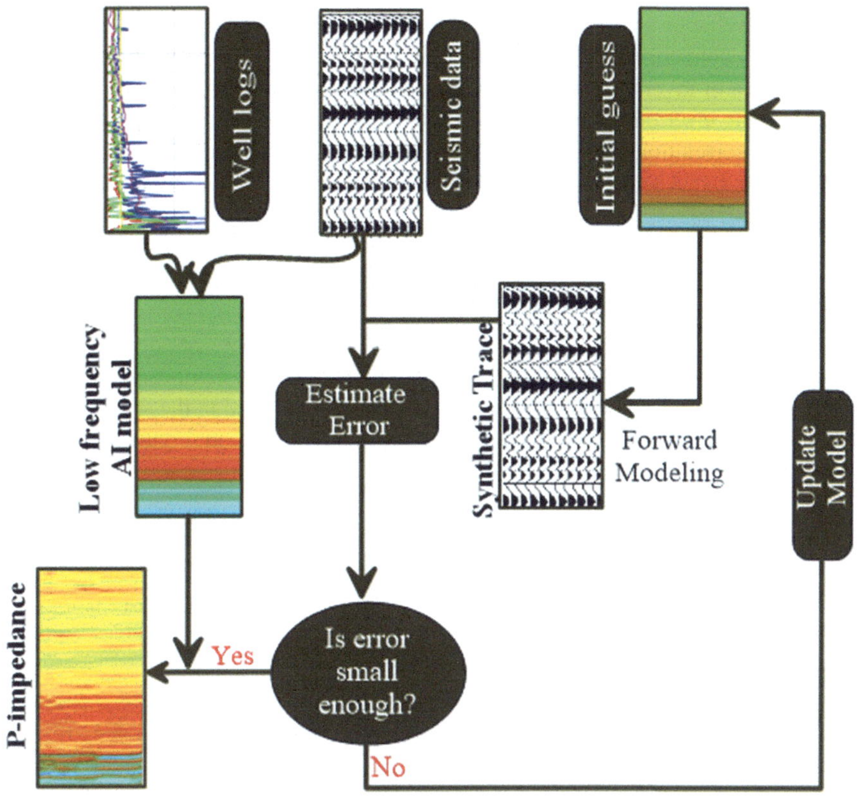

Fig. 3.15 Flowchart of Model-based inversion method

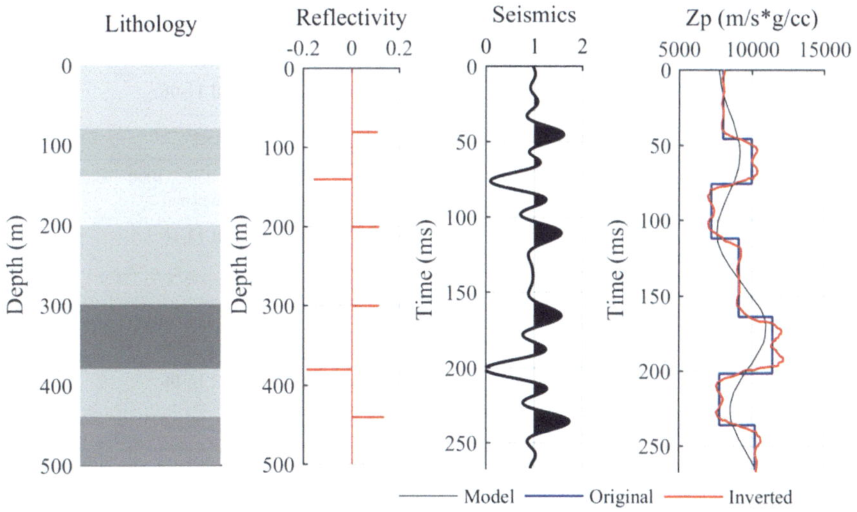

Fig. 3.16 Synthetic example for model-based inversion methods

from given velocity and density combination, track 3 shows synthetic seismogram generated by convolution of 30 Hz wavelet with reflectivity series and track 4 shows a comparison of inverted impedance with true impedance. From the figure, one can notice that the inverted impedance is following the trend of the true impedance very nicely. The correlation coefficient between inverted and true impedance is 0.99 and RMS error is 0.023 which indicates the good performance of the algorithm.

3.5.4 Application of MBI to Real Data

The real data example is applied in two steps, in the first step, composite traces near to well locations are extracted and MBI method is applied and in the second step, it is applied to the 3D data volume to get AI in the inter-well region. Figure 3.17 shows

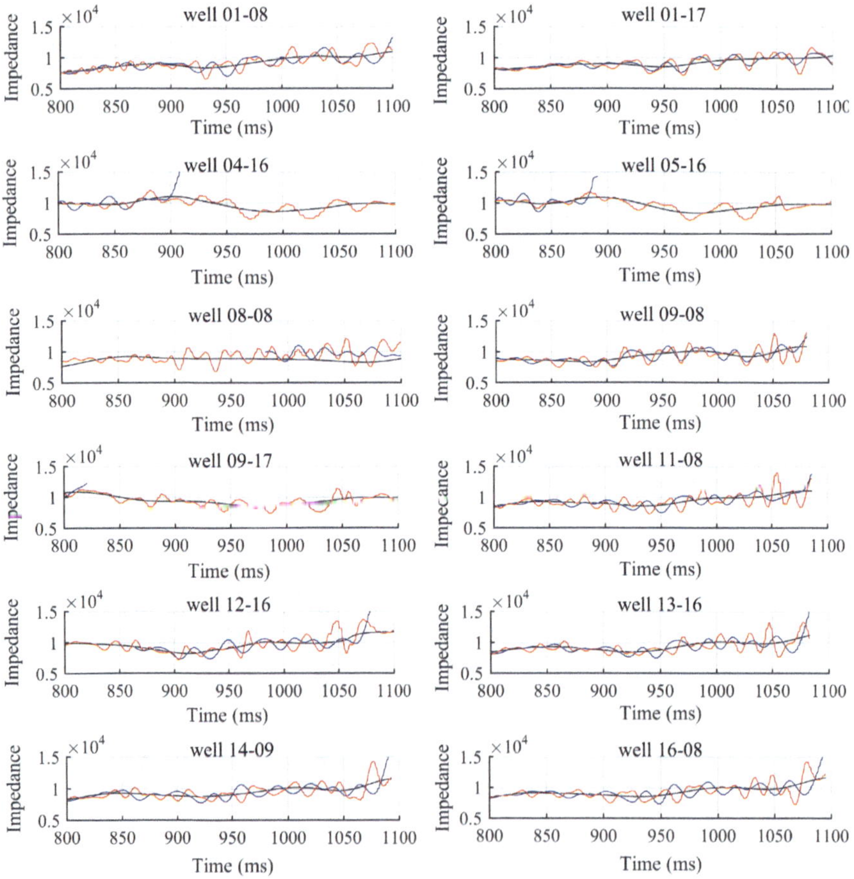

Fig. 3.17 Inversion analyses for composite trace using the model-based inversion method

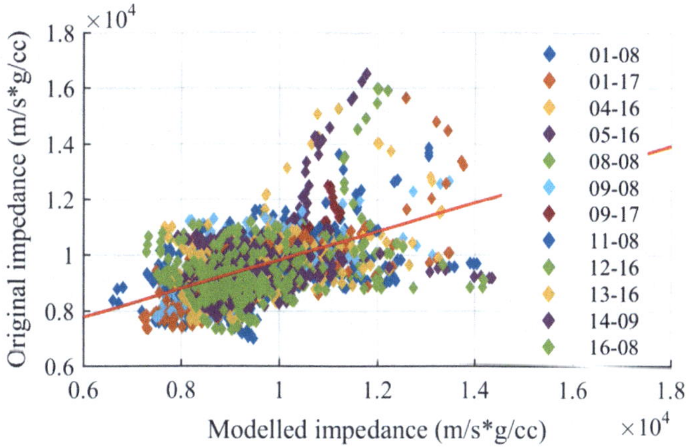

Fig. 3.18 Crossplot of inverted impedance for composite trace

inversion results for composite trace. The figure shows 12 subplots each corresponds to 12 different well logs. Each subplot has three curves, the black curves correspond for the initial guess model, the blue curve indicates true AI from logs and the red curve indicates inverted AI. The analysis of composite trace shows that the inverted results are following the trend of original AI very nicely for all wells. The analysis of composite trace is very important to cross verify inversion results as one knows true AI from the well log data.

A cross plot between inverted and original AI has been generated for all the wells and is shown in Fig. 3.18. Next, a best fit straight line is fitted to the data. The deviation from this best fit line shows the quality of the inversion results. As if all the points lying on the fitted line indicates that the inverted results are the same as original data and if shows small deviation means small difference in between original and inverted data points. From Fig. 3.18, one can notice that most of the data points lie near to the best fit line and demonstrate good results.

The variation of correlation coefficient and RMS errors are shown in Fig. 3.19. The correlation coefficient and RMS errors are very fundamental parameters to classify the difference between two similar signals. The figure shows the variation of correlation coefficient (top) and RMS error (bottom) with well logs. From the figure, one can notice that the correlation coefficient is very high and RMS error is very low which is obvious. The correlation coefficients vary from 0.997 to 0.995 and the average value is 0.99. On the other hand, the RMS error shows variation from 700 to 1100 (m/s)*(g/cc) with average value of 920 (m/s)*(g/cc). The other statistical parameters are shown in Table 3.3. These values indicate the high performance of the MBI algorithm.

We have seen that the inverted results are very accurate and give satisfactory results for the composite trace and now we are in a position to apply MBI methods to the real 3D seismic volume to estimate acoustic impedance volume. The parameters used for model-based inversion are as follows.

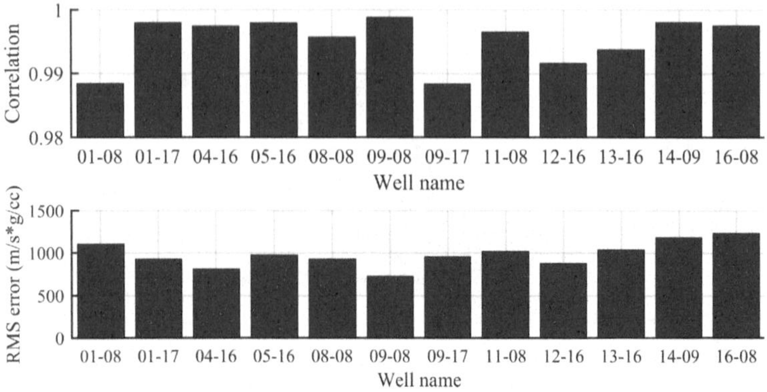

Fig. 3.19 Variation of correlation (top) and RMS error (bottom) for results estimated MBI

Table 3.3 Quantitative comparison of seismic inversion results (AI stands for acoustic impedance, CC stands for colored inversion, PSNR stands for peak signal to noise ratio, MAE stands for mean absolute error and SAE stands for sum absolute error)

Model-based inversion						
Properties	Reservoir zone	CC	RMS Error	PSNR	MAE	SAE
AI	$6.3 \times 10^3 - 8.0 \times 10^3$	0.86	2×10^3	17.9	1.6×10^3	1.1×10^5
Amp. spectrum	NA	0.99	0.0	80.1	0.0	2.7

- Inversion time interval: 300–1100 ms;
- Number of iterations: 20;
- Separate scales;
- Impedance change constraint: ±30 percent;
- High cut frequency: 12 Hz.

Figure 3.20 depicts a cross-section of inverted acoustic impedance at inline 28. The figure shows a very high resolution of subsurface layers as compared with seismic data which shows only interface property. This section makes an easy interpretation of seismic data. To cross verify inverted results, the AI from well 08-08 is also shown above the inverted section and found that both are matching very nicely. One can also notice that the area has a variety of acoustic impedance from 6000 16,000 m/s*g/cc. The acoustic impedance varies vertically and laterally which can also detect faulting plane, anticline, syncline, etc. in the subsurface. The inverted section detects low impedance anomaly near to 1060–1075 ms time interval which may correspond to sand channel. This low impedance zone is also visible in well log data.

Thereafter, to see the variation of the frequency band in the inverted section, one can compare the amplitude spectrum of an inverted synthetic with the amplitude spectrum of the input seismic. Figure 3.21 shows comparison of the amplitude spectrum of inverted synthetic and input seismic data. Visually, one can notice that the

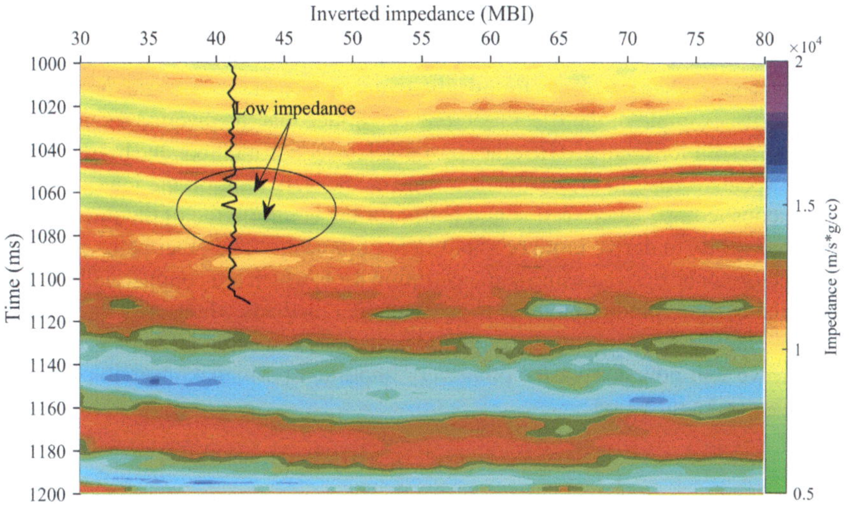

Fig. 3.20 Cross-section of inverted acoustic impedance (inline 41) estimated using the MBI technique

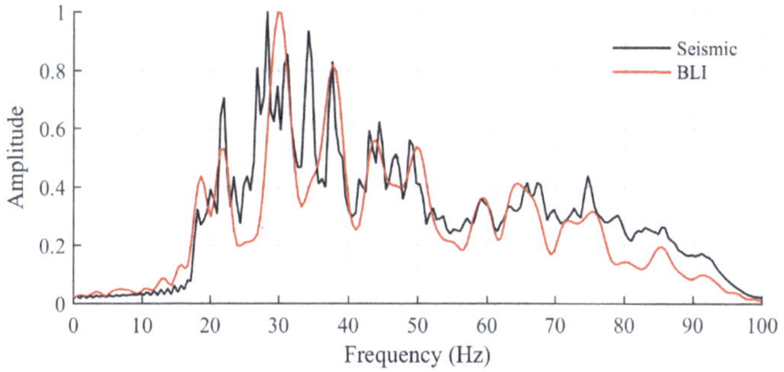

Fig. 3.21 Comparison of the Amplitude spectrum of inverted synthetic and Blackfoot seismic before application of MBI methods

amplitude spectrum of inverted synthetic is almost matching with the amplitude spectrum of Blackfoot seismic. This is the biggest advantage of using the MBI method over other inversion methods. We have seen that the BLI and CI distort frequency content but here the MBI methods have not much destroyed the frequency of the input signal. This is obvious because the MBI method does not use seismic data directly as input whereas the other inversion method uses seismic data directly as input and mathematical process distort frequency content.

The model-based inversion method has a comparatively complex mathematical background but can be easily used to the real field data. The MBI method is the most popular post-stack seismic inversion method and the results derived from it can be used for the quantitative interpretation. The non-uniqueness is a problem of the entire inversion methods hence one uses a low-frequency model to constraint inversion results. Comparatively, the model-based inversion methods are time taking hence it is used when one seeks for quantitative variation of the acoustic impedance. In the above example of the Blackfoot, one can notice that the inverted section have very high resolution compared to other post-stack inversion methods and help to interpret prospective zone in the subsurface. One cannot rely only on the interpretation of the acoustic impedance section and hence for the confirmation of the prospective zone, one needs to derive more petrophysical parameters. The other petro-physical parameters can be derived using some perdition tools which are commercially available. One of these prediction tools is geostatistical methods which are largely used in the geophysics community. This method is discussed in detail in chapter 8. In this chapter, we will now discuss the other very important post-stack inversion method i.e. sparse spike inversion.

3.6 Sparse Spike Inversion

The sparse spike inversion methods is another type of post-stack inversion methods developed in the 1980s (Zhang et al. 2016) but now has become the common method to estimate petrophysical parameters. There is an assumption in sparse spike inversion that the earth's reflectivity is composed of large spikes with small spikes in the background. Lithological, these large spikes are meaningful which corresponds to unconformity and hence the sparse spike method tries to find these large spikes and ignore small spikes (Bosch et al. 2010; Maurya and Singh 2015; Maurya and Sarkar 2016). The process is similar to the model-based inversion methods where one assumes simple earth's reflectivity model and calculate synthetic seismic trace by the convolution of extracted statistical wavelet and thereafter, the error is estimated between synthetic seismic and seismic data which can be minimized by adding more number of spikes in the reflectivity series (Debeye and Riel 1990). When the sparse spike inversion is constrained by a low-frequency model derived from acoustic impedance well logs or geologic models, it is commonly referred to as model-based inversion (Russell 1988).

The purpose of sparse spike impedance inversion is to obtain the acoustic impedance volume from seismic data. The inverted impedance possesses broadband of spectrum, displays a blocky structure of the subsurface in the time domain, and is directly related to the lithology. The bandwidth of seismic data is generally 10–80 Hz frequency, so lack of high and low-frequency content. To get a broadband spectrum of the inverted impedance volume, additional information is needed which generally comes from well log data. The Sparse Spike solution does not have any low-frequency components due to the band-limited nature of input seismic. Therefore, the low-frequency trend of the model is imported from the initial model. The recursive method uses a feedback mechanism to generate a more satisfactory output. The inversion solution may vary considerably from trace to trace, thus making the reliability of the output weaker. A low frequency AI variation trend can be imported to obtain more appropriate results and to get a better convergence for the found solutions from trace to trace. The constrained option uses a low-frequency model as a guide. The low-frequency variation is estimated from filtered well logs and this gives much better results. The inversion replaces the seismic trace by a pseudo acoustic impedance trace at each CDP position.

A flow chart of the sparse spike inversion method is presented in Fig. 3.22 which includes both inversions. The sparse spike inversion methods are divided into two broad categories on the basis of the minimization of error. The first method is called Linear Programming inversion (LPI) that uses a l_1-norm solution for its implementation and the second method is called Maximum Likelihood inversion (MLI) which uses l_2-norm solution for its implementation (Russell 1988; Sacchi and Ulrych 1995; Zhang and Castagna 2011). These methods are briefly explained in the following sections.

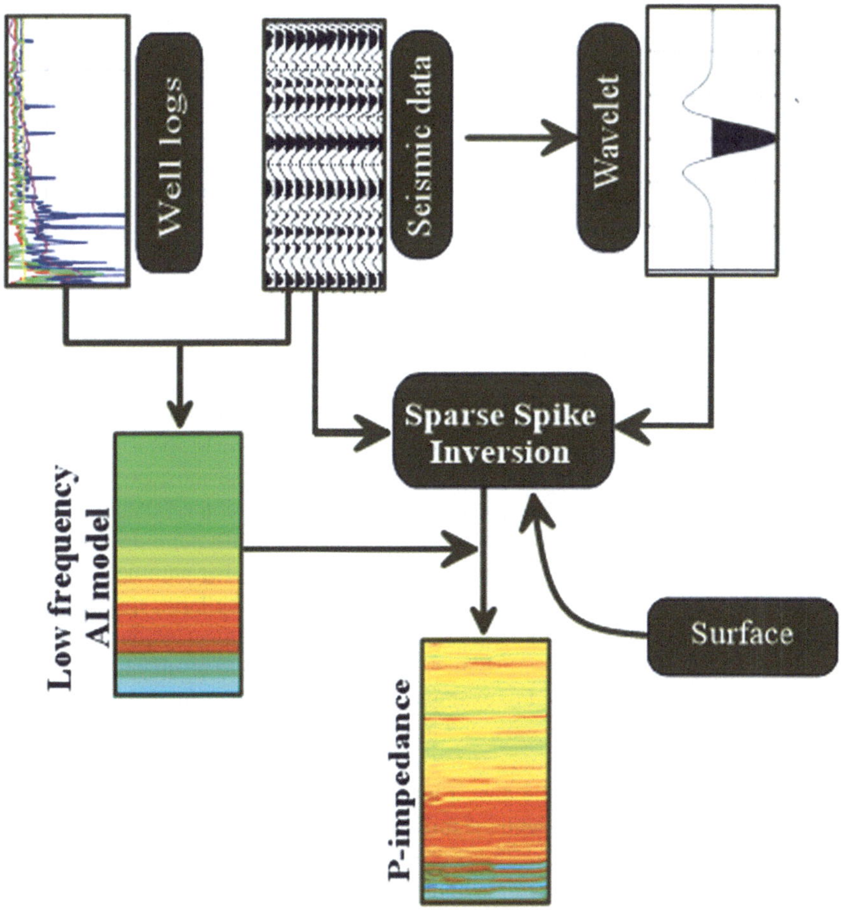

Fig. 3.22 Flow chart of sparse spike inversion method

3.6.1 Maximum Likelihood Inversion (MLI)

The maximum likelihood inversion first performs the maximum likelihood deconvolution to estimate the earth's reflectivity and then it is transformed into the acoustic impedance under the maximum likelihood inversion scheme. The fundamental assumption of Maximum-Likelihood deconvolution is the same as the sparse spike inversion methods which state that the earth's reflectivity is composed of a series of large spikes superimposed over small spikes in the background. This contrasts with the spiking deconvolution, which assumes a perfectly random distribution of reflection coefficients.

From the above assumptions about the model parameter, one can derive an objective function that may be minimized to yield the optimum or most likely reflectivity (Kormylo and Mendel 1983; Mendel 2012). It is noteworthy that this method gives an estimate of both the sparse reflectivity and wavelet.

The objective function can be written for reflectivity and noise which may be minimized to produce the reflectivity and wavelet combination consistent with the statistical assumption (Russell 1988; Li and Speed 2004; Maurya and Singh 2018). The objective function E is given by

$$E = \sum_{k=1}^{L} \frac{r_k^2}{R^2} + \sum_{k=1}^{L} \frac{n_k^2}{N^2} - 2m\ln(\lambda) - 2(L - m)\ln(\lambda) \quad (3.29)$$

where r_k is the reflection coefficient at the kth interface, m is the total number of reflections, L is the total number of samples presents, N is the square root of noise variance, n_k is noise at the kth sample, and λ is the likelihood that a given sample has a reflection.

Mathematically, the expected behavior of the objective function is expressed in terms of the parameters shown above. No assumptions are made about the wavelet. The reflectivity sequence is postulated to be sparse, meaning that the expected number of spikes is governed by the parameter lambda (λ), the ratio of the expected number of nonzero spikes to the total number of trace samples. Normally, lambda is a number much smaller than one (Hampson and Russell 1985). The other parameters needed to describe the expected behavior are R, the RMS size of the large spikes, and N, the RMS size of the noise. With these parameters specified, any given deconvolution solution can be examined to see whether it is likely to be the result of a statistical process with those parameters. For example, if the reflectivity estimate has a number of spikes much larger than the expected number, then it is an unlikely result.

In simpler terms, we are looking for a solution with the minimum number of spikes in its reflectivity and the lowest noise component. It is noteworthy that the objective function for the one with the minimum spike structure is indeed the lowest value.

Of course, there may be an infinite number of possible solutions, and it would take too much computer time to look at each one. Therefore, a simpler method is used to arrive at the answer. Essentially, one starts with an initial wavelet estimate of the sparse reflectivity, improves the wavelet and iterate through this sequence of steps until an acceptably low objective function is reached (Mendel 2012). Thus, there is a two-step procedure- having the wavelet estimate, update the reflectivity, and then, having the reflectivity estimate, update the wavelet.

After getting reflectivity from the maximum likelihood deconvolution, it is transformed into the acoustic impedance which is more meaningful in comparison to the reflectivity series (Goutsias and Mendel 1986). This conversion can be performed in two ways; the first approach is that the acoustic impedance can be estimated by using direct relationship between reflectivity and impedance (Velis 2006, 2007; Maurya and Sarkar 2016).

$$Z_{j+1} = Z_j \prod_{k=1}^{j} \left(\frac{1 + r_k}{1 - r_k} \right) \tag{3.30}$$

But this relationship has a limitation as if the data contains significant noise then the transformation cannot be performed properly as the relationship does not include any noise component. A second approach is used to overcome the problem and to generate an acoustic impedance model from the reflectivity section which included noise component also and hence gives complete spectrum (Russell 1988; Wang et al. 2006; Zhang et al. 2016). The equation is as follows.

$$\ln Z(i) = 2H(i) * r(i) + n(i) \tag{3.31}$$

where $Z(i)$ is the known impedance trend, $r(i)$ is earth reflectivity series and $n(i)$ is the noise in the input trend and $H(i)$ is a factor that would be given by

$$H(i) = \begin{cases} 1, i < 0 \\ 0, i > 0 \end{cases} \tag{3.32}$$

Let us look for an example of the maximum likelihood inversion method. The example is provided for the real data from the Blackfoot field, Alberta, Canada.

i. **Application of maximum likelihood inversion**

The maximum likelihood inversion is performed to estimate acoustic impedance from seismic reflection data with the integration of well log data. Similar to the other inversion methods, the MLI is performed in two steps, first, the inversion of composite trace near to well locations and second, entire seismic volume inversion. Figure 3.23 depicts the comparison of inverted acoustic impedance with well log impedance for 12 wells. In Figure, the black solid color shows the initial guess model, the blue solid curve depicts well log acoustic impedance and inverted acoustic impedances are shown in red solid color. From Fig. 3.23, it is noticed that the inverted impedances are agreeing very well with the well log impedances. The well log data is not recorded from the surface to bottom in all wells and hence in some wells, the data is available between 1000 and 1100 ms time intervals only.

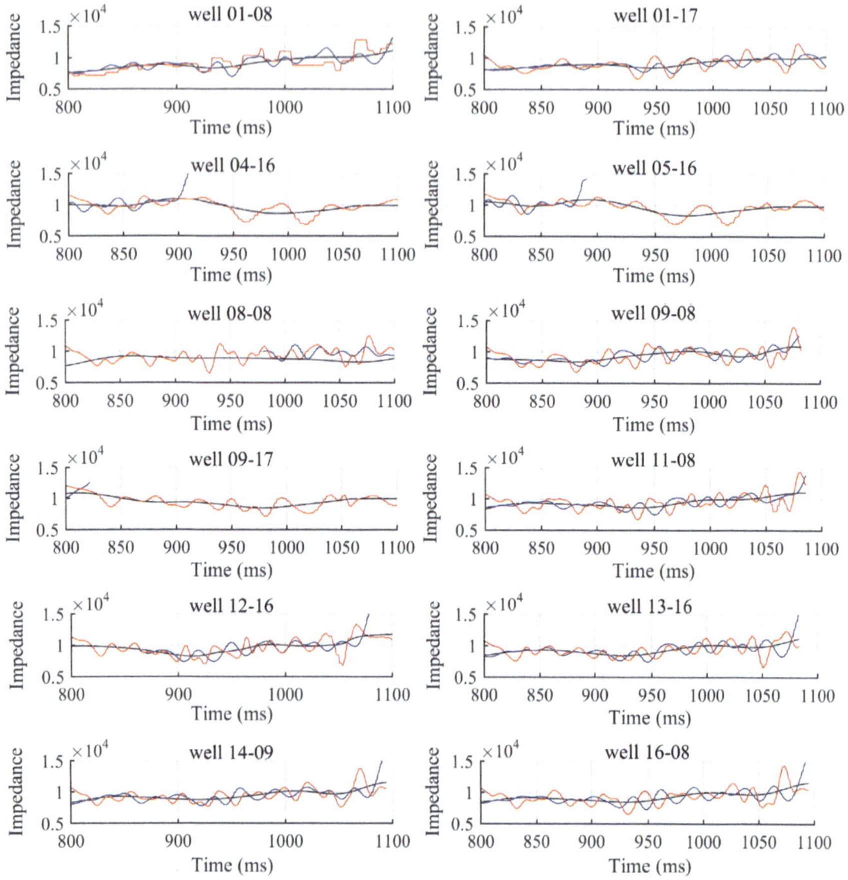

Fig. 3.23 Inversion analyses for the composite trace near to well locations for maximum likelihood inversion methods

Thereafter, for a quality check of inverted results, a cross plot has been generated of the inverted impedance and actual impedance for all thirteen wells and displayed in Fig. 3.24. The distributions of scatter points show the quality check of inversion results and hence performance of the maximum likelihood inversion. It is noticed that the inverted data points are very close to the best fit curve which represents the sample by sample comparison of inverted result with the original one and hence depicts the good performance of the MLI algorithm.

Till now the results are cross-verified qualitatively but now the analysis will be extended for quantitative comparison so that one can exactly know what is the difference between inverted and original results.

For the composite data case, we have an estimated correlation coefficient and RMS Error which are presented in Fig. 3.25. The figure has two curves, the top figure shows the variation of correlation coefficient whereas the bottom figure depicts RMS

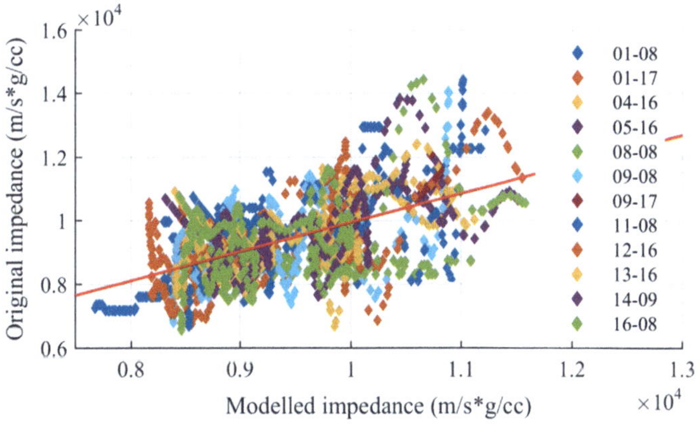

Fig. 3.24 Crossplot of inverted acoustic impedance for composite trace versus original impedance from well logs

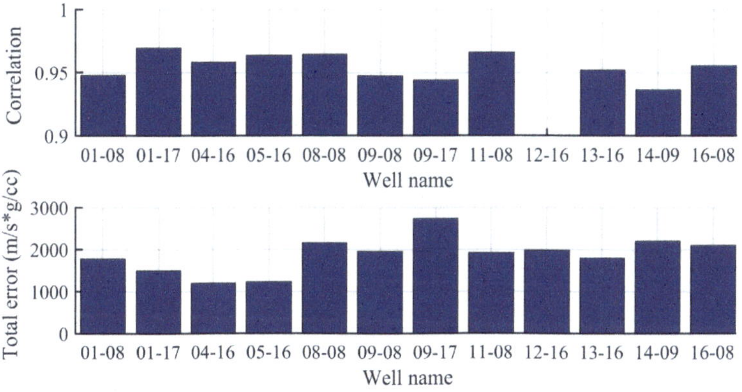

Fig. 3.25 Variation of correlation coefficient (top) and RMS error (bottom) for inverted results of the composite trace

Error for wells. It is noticed that the correlation coefficients varies from 0.96 to 0.98 with mean value of 0.975 and RMS error varies from 1000 to 2100 m/s*g/cc with mean value of 1300 m/s*g/cc. These values indicate that the algorithm successfully searched the optimum solution.

Till now the analysis is limited at one point but one seeks to see the variation of AI in the subsurface vertically as well as horizontally in the subsurface. Hence, the maximum likelihood inversion is applied to the 3D seismic volume to estimate AI in the inter-well region. The data used is from the Blackfoot field, Canada. The seismic section is inverted for acoustic impedance and is shown in Fig. 3.25 at inline 28. The special features are highlighted in the inverted impedance section. The inverted curves show a very high-resolution image of the subsurface. The well log impedance is also shown above it and shows nice matching between them. The variation of acoustic impedance of the area is 6000–16,000 m/s*g/cc. The analysis of the inverted impedance section also shows a low impedance anomaly ranging from 6000 to 9000 m/s*g/cc within 1060–1075 ms time interval.

Frequency comparison of the amplitude spectrum of the input seismic and inverted synthetic seismic data is displayed in Fig. 3.26. From the figure, it is noticed that the amplitude spectrum of both the data is matching very nicely and hence one can conclude that the maximum likelihood inversion method keeps all the frequency content as in MBI methods. This is great advantage of MLI methods over other inversion methods. The figure also depicts that for lesser frequency, the amplitude is matching accurately with the seismic amplitude whereas for the larger frequency (>80 Hz) the amplitude spectrum shows small deviation from the Blackfoot seismic amplitude. This is due to the fact that the MLI algorithm distorts the amplitude spectrum, particularly for larger frequency during its implementations (Fig. 3.27).

A time slice includes simultaneously level events from more than one reflection horizon. A high-frequency spatial event on a slice of time is either a steep dipping event or a timed event of high frequency. Thus, a high dip from the event's high-frequency character can be inferred from the time slices and smooth drop from the low-frequency character. In addition, time slices can trace contours connected with a reflection horizon. If the contours between a shallow and a deeper slice of time are narrow, then the character is a low structure. By contrast, if the contours widen between a slice of shallow and a slice of deeper time, the feature is a high structural.

A slice of AI at 1065 ms time interval is generated and shown in Fig. 3.28. This slice shows the horizontal variation of the acoustic impedance in the subsurface. The slice generation is again a very important step to represent horizontal variation of the AI and hence horizontal variation of reservoir zone if any. Figure 3.28 highlights the

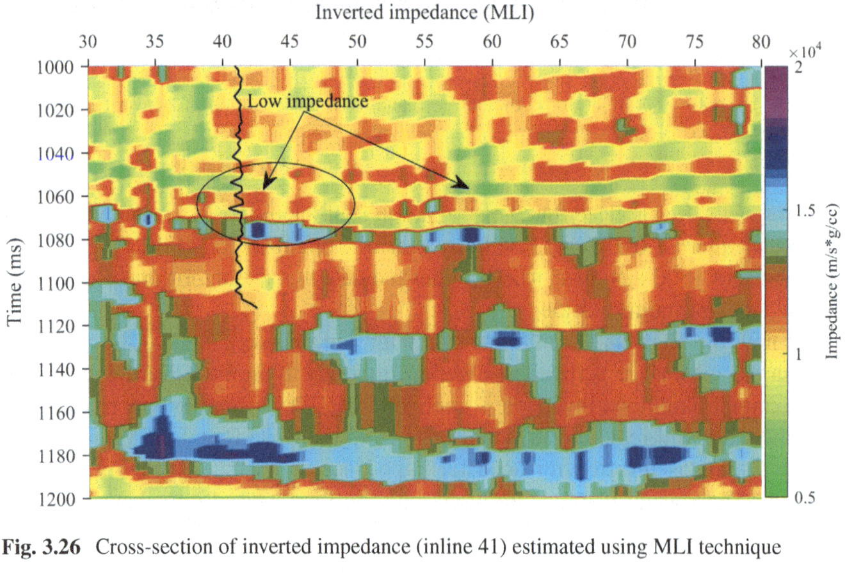

Fig. 3.26 Cross-section of inverted impedance (inline 41) estimated using MLI technique

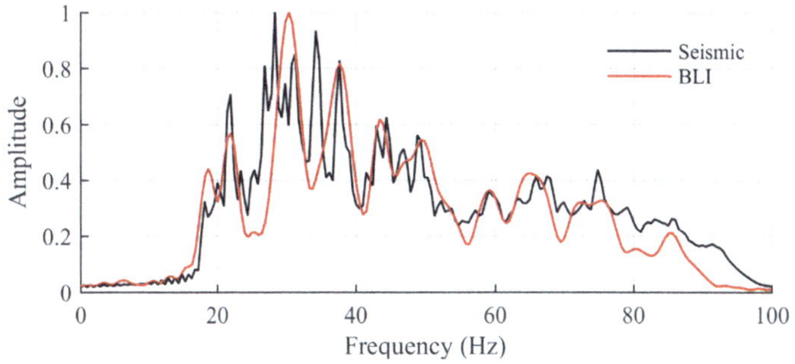

Fig. 3.27 Comparison of the amplitude spectrum of Blackfoot seismic and inverted synthetic seismic

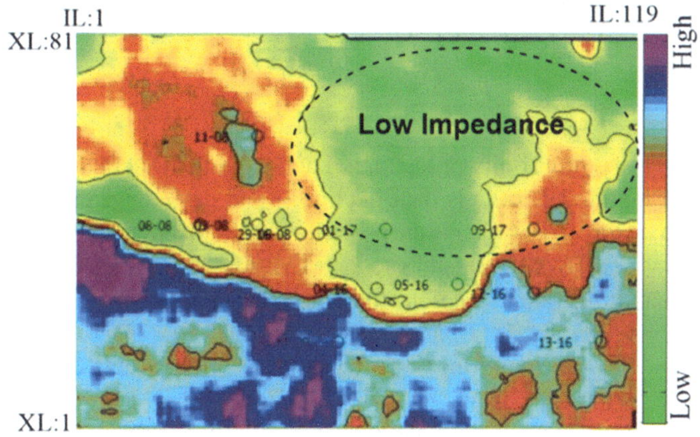

Fig. 3.28 Impedance slice at 1065 ms time interval

expected reservoir zone horizontally which is predicted in the vertical AI section from almost all the inversion results discussed above. It is also noticed that the reservoir zone varies from the SW direction to the NE direction.

3.6.2 Linear Programming Inversion (LPI)

The Linear Programming inversion method is another type of sparse spike inversion technique available to estimate the earth's reflectivity series that approximates the seismic data with minimum number of spikes (Li 2001; Russell and Hampson 1991; Maurya and Singh 2015). The L_1 norms (Linear Programming) logarithms are used by the LPI method to estimate the acoustic impedance of the subsurface (Loris et al. 2007; Zhang et al. 2016). The LPI method tends to remove the wavelet effect from the seismic data and hence increases the band of inversion results which maximizes vertical as well as the horizontal resolution of the data and also minimizes the tuning effects (Helgesen et al. 2000).

The mathematical background of the Linear Programming sparse spike inversion is developed by Qing Li in (Li 2001; Sacchi and Ulrych 1995, 1996). The aim of the seismic inversion technique is to find the subsurface model that is reflection coefficient $r(t)$ in this study, given that seismic data $s(t)$ and source wavelet $w(t)$ as input, and then calculate acoustic impedance $Z(t)$. There is a relationship between data $d = (x1; x2; \ldots)$, model parameters $m = (r1; r2; \ldots)$ and noise n as following.

$$Lm + n = d \tag{3.33}$$

where L is operator. Generally, the observation d is known, and then the subsurface model can be defined by the probability $p(m|d)$ (Li 2001; Sacchi and Ulrych 1995). This probability can be described by the Bayes formula in the following way.

$$p(m|d) = \frac{p(d|m)p(m)}{p(d)} \tag{3.34}$$

where $p(d|m)$ is the probability or likelihood of obtaining the data, $p(m)$ is the prior probability of the model, $p(d)$ is data likelihood and enters into the problem as a normalization factor, and $p(m|d)$ is the posterior probability of the model (Barrodale and Roberts 1973, 1978). One could use the MAP solution, m_{MAP} that maximize a posterior probability $p(m|d)$. An objective function can be chosen as follows.

$$E = -\log(p(m|d) = \log(p(d|m)) - \log(p(m)) \tag{3.35}$$

In Eq. 3.35, $p(d)$ is a constant and hence omitted for the simplicity. The prior knowledge of the model is generally given as a global constraint $S(m)$. Thereafter, one can use the principle of maximum entropy to compute the prior probability (Debeye and Riel 1990; Sacchi and Ulrych 1995).

Consider continuous model parameters m with probability density function $p(m)$. The entropy h is given by the following equation.

$$h = \int p(m)\log[p(m)]dm \tag{3.36}$$

Equation 3.36 expresses the uncertainty associated with the distribution $p(m)$. If the information about m is available in the form of some global constraint $S(m)$ then, the corresponding maximum entropy probability distribution will be given as follows (Russell 1988; Oliveira and Lupinacci 2013):

$$p(m) = Ae^{-S(m)} \tag{3.37}$$

where A is normalization constant. The generalized version of $p(d|m)$ can be written as a function of discrepancy between the model and the observation as following.

$$p(d|m) = \frac{p^{1-\frac{1}{p}}}{2\sigma_p \Gamma\left(\frac{1}{p}\right)} exp\left(\frac{-1}{p}\frac{|d - Lm|^p}{(\sigma_p)^p}\right) \tag{3.38}$$

Based on Eq. 3.38, one can choose the objective function with the different norms to solve the inverse problem. In this way, one can maximize the objective function E and can obtain a solution that gives the least error between the model parameters and the observations (Li 2001; Oliveira and Lupinacci 2013; Maurya and Singh 2018). The solution of the l_1- norm can be estimated as follows.

Consider the convolution theory which states that the seismograms $s(t)$ are generated by the convolution of earth's reflectivity series $r(t)$ with a known wavelet $w(t)$, if the data does not contain type of noises (Velis 2006, 2007; Wang 2010; Maurya and Sarkar 2016).

$$S(t) = r(t) * w(t) \tag{3.39}$$

For the layered earth, the reflectivity function will be zero everywhere except at those times corresponding to the two-way travel time to the jth layer. Thus, the reflectivity function has the following mathematical form (Oldenburg et al. 1983; Zhang and Castagna 2011).

$$r(t) = \sum_{j=1}^{NL} r_j \delta\left(t - \tau_j\right) \tag{3.40}$$

where NL is the total number of layers in the earth subsurface model, r_j is the reflectivity coefficient at the jth and (j + 1)th interface. Generally, the total number of data points in the seismogram i.e. N is much larger than the number of layers i.e. NL in the models hence; it greatly reduces the degree of freedom of the model and hence reduces the non-uniqueness (Brien et al. 1994; Wang 2010).

Further, recall the acoustic impedance equation and which for the jth layer is defined as follows.

$$Z_j = \rho_j v_j \tag{3.41}$$

where ρ_j is density and v_j is the velocity of the jth layer. The earth reflectivity can be estimated from the impedance model as following.

$$r_j = \frac{Z_{j+1} - Z_j}{Z_{j+1} + Z_j} \tag{3.42}$$

A rearrangement of Eq. 3.42 gives the following relation.

$$Z_{j+1} = Z_j \frac{1 + r_j}{1 - r_j} = Z_1 \prod_{j=1}^{k}\left(\frac{1 + r_j}{1 - r_j}\right) \tag{3.43}$$

Equation 3.43 shows how the impedance is related to the reflectivity coefficients. The reflectivity coefficients are related to acoustic impedance for continuous travel time as follows.

$$r(t) = \frac{1}{2}\frac{d[\ln Z(t)]}{dt} \tag{3.44}$$

In deriving Eq. 3.44, second-order quantities have been omitted because they are almost negligible. Equation 3.44 may be rewritten as follows.

$$Z(t) = Z(0)\exp\left[2\int_0^t r(t)dt\right] \tag{3.45}$$

From the above relation, we may see that the sparseness in $r(t)$ may result in a blocky structure in Z(t) (Oldenburg et al. 1983).

i. **Application of linear programming inversion**

The linear programming inversion method is applied to the post-stack seismic data from the Blackfoot field, Canada. As the other inversion methods, the LPI also integrates seismically and well log data together. These two data (Seismic data and well logs) are generally in two different domains, one is in time (seismic) and the other is in-depth (well log). Hence, before proceeding towards the inversion, one needs to convert them in the same domain. After conversion, the other processes of input preparation i.e. generation of initial guess model, extraction of statistical wavelet, etc. have been performed. These processes are already explained in Chap. 3. After setting all the input, Linear Programming inversion is performed to the data in two steps like other inversion methods. Firstly, the inversion is applied to the extracted composite trace near to well locations and secondly, the entire seismic volume is inverted for AI.

Figure 3.29 shows inverted results for composite traces near to well locations along with actual impedance from well log data. The Figure contains three curves in one plot, the black solid line corresponds to the initial guess model, the blue solid curve represents a well log AI curve and the red solid curve represents inverted AI. From the figure, one can notice that the inverted curves are following the trend of the original curve very nicely. Here, one can also notice that the phase of the inverted curve is also matching with the phase of the original curve whereas the other inversion methods show a lack of phase matching.

Further, a cross plot of inverted AI with original AI from the wells is generated and shown in Fig. 3.30. Thereafter, a best fit straight line is fitted to the data to represent trend of scatter points. Now, one can notice, the variation of scatter points from the best-fitted line, if the points lie close to the best fit line then it indicates the good conversion of seismic amplitude into the acoustic impedance as the inverted curves do not deviate much from the original one and vice versa. Figure 3.46 indicates the variation of scatter points very close to the best fit line for almost all thirteen wells and hence represents the good performance of the algorithm.

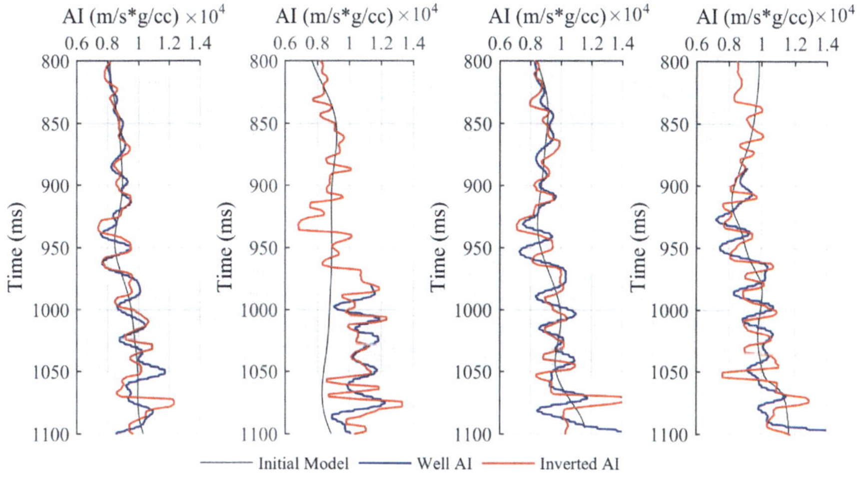

Fig. 3.29 Inversion analysis for composite trace near to well locations. The results for the four wells are shown here

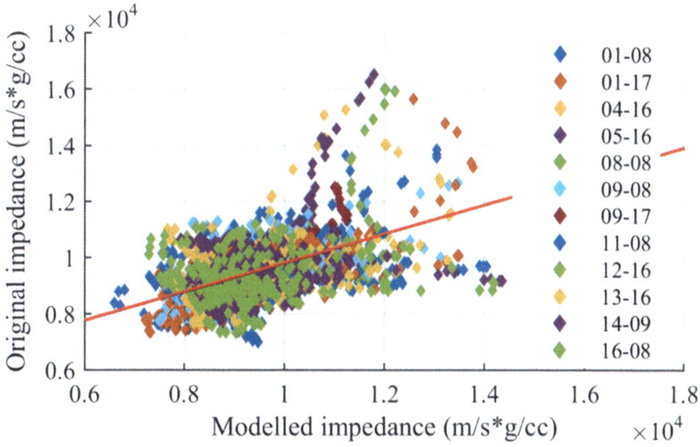

Fig. 3.30 Crossplot of inverted impedance from the composite trace versus original impedance from well logs

Variation of correlation coefficient and RMS error between original and inverted acoustic impedance are shown in Fig. 3.31. The top histogram represents the correlation coefficient and the bottom histogram represents RMS error with wells. It can be notice that the correlation varies from 0.987 to 0.999 with mean value of 0.993 and RMS error varies from 700 to 1400 m/s*g/cc with mean value of 950 m/s*g/cc. These values indicate very high performance of the algorithm and represent a very high quality of the inversion results.

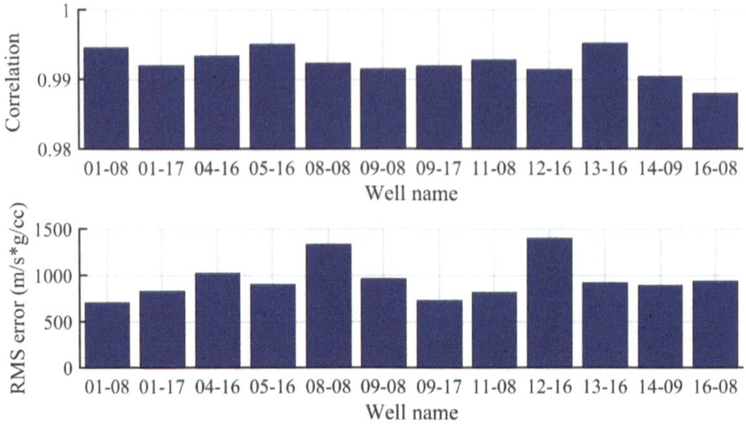

Fig. 3.31 Variation of correlation coefficient (top) and RMS error (bottom) with wells

After getting satisfactory results for the analyses of the composite traces, now, one can perform inversion of entire seismic volume. Figure 3.32 depicts a cross-section of inverted impedance at inline 28. The figure shows variation of acoustic impedance with different colors. The inverted section shows a higher resolution image of the subsurface compare to the seismic section and highlights the thinner reflector more clearly. It is noticed that the area has a variety of acoustic impedance from 5000 the 20,000 m/s*g/cc. The analysis of the inverted section also shows a low impedance anomaly ranging from 6000 to 8500 m/s*g/cc in between 1060 and 1075 ms two

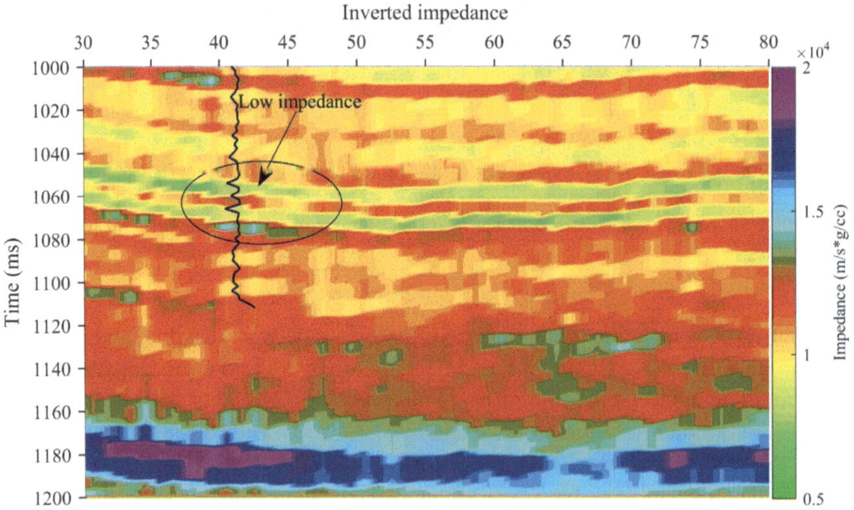

Fig. 3.32 Cross-section of inverted impedance (inline 28) estimated using LPI methods

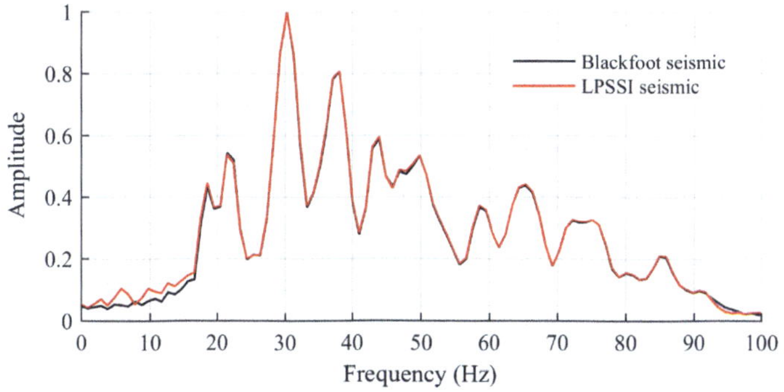

Fig. 3.33 Comparison of amplitude spectrum between Blackfoot seismic and inverted synthetic seismic

way travel time. This low AI zone is interpreted as sand channel zone although this is prelim interpretation and for the confirmation of this channel needs more parameters.

The frequency content of the inverted section becomes degraded as the inversion involves mathematical transformation which sometimes degrades frequency contents hence needs to analyze. A comparison of the amplitude spectrum between inverted synthetic seismic and Blackfoot seismic is shown in Fig. 3.33. The figure depicts that both the spectrum follow each other very well and indicates that the algorithm preserves all the frequency contents during its implementations. One more thing that can be noticed here is that the amplitude spectrums of the inverted curves are matching exactly to the input seismic which is not in the case of the other inversion methods. This are great advantages of the LPI methods. Although sometimes the degradation of the frequency content of the inverted section also depends on the choice of appropriate parameters during the inversion process. If one does not choose inversion parameters according to the input data, then the amplitude spectrum will not match exactly.

The slice at 1065 ms time interval is displayed in Fig. 3.34. A horizontal display or map view of 3D seismic data having a certain arrival time, as opposed to a horizon slice that shows a particular reflection. A time slice is a quick, convenient way to evaluate changes in the amplitude of seismic data. The variations of acoustic impedance along horizontal are displayed in the figure. The interpreted sand channel is also highlighted by the ellipse. It is noticed that the sand channel varies from NE to SW direction.

As the seismic data shows amplitude variation along with time and cannot be interpreted in the form of subsurface lithology whereas by using these inversion techniques one can obtain the distribution of petrophysical parameters in the inter-well region. This is very helpful to interpret subsurface lithology particularly in exploration and production projects.

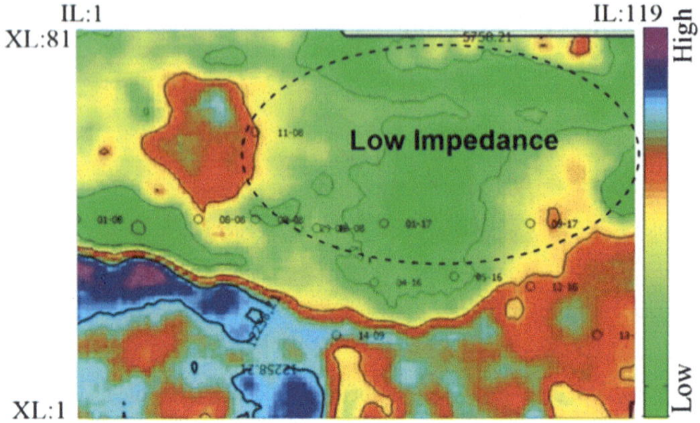

Fig. 3.34 Impedance slice at 1065 ms time interval

References

Ansari HR (2014) Use seismic colored inversion and power-law committee machines based on imperial competitive algorithm for improving porosity prediction in a heterogeneous reservoir. J Appl Geophys 108:61–68

Barrodale I, Roberts FD (1973) An improved algorithm for discrete l-1 linear approximation. SIAM J Numer Anal 10(5):839–848

Barrodale I, Roberts F (1978) An efficient algorithm for discrete l_1 linear approximation with linear constraints. SIAM J Numer Anal 15(3):603–611

Bosch M, Mukerji T, Gonzalez EF (2010) Seismic inversion for reservoir properties combining statistical rock physics and geostatistics: a review. Geophysics 75(5):75A165–75A176

Brien OM, Sinclair AN, Kramer SM (1994) Recovery of a sparse spike time series by l/sub 1/norm deconvolution: IEEE Trans Signal Process 42:3353–3365

Brossier R, Operto S, Virieux J (2015) Velocity model building from seismic reflection data by full-waveform inversion. Geophys Prospect 63(2):354–367

Brown AR (2004) Interpretation of three-dimensional seismic data. AAPG Memoir 42. SEG Investigation in Geophysics, No. 9. AAPG, Tulsa

Clochard V, Delépine N, Labat K, Ricarte P (2009) Post-stack versus pre-stack stratigraphic inversion for CO_2 monitoring purposes: a case study for the saline aquifer of the Sleipner field. SEG Annual Meeting, Society of Exploration Geophysicists

Cooke DA, Schneider WA (1983) Generalized linear inversion of reflection seismic data. Geophysics 48(6):665–676

Debeye H, Riel VP (1990) Lp-norm deconvolution: geophysical Prospecting 38:381–403

Ferguson RJ (1996) PS seismic inversion: modeling, processing and field examples. M.Sc. Thesis, University of Calgary, Canada

Ferguson RJ, Margrave GF (1996) A simple algorithm for band-limited impedance inversion. CREWES Res Rep 8(21):1–10

Goutsias J, Mendel JM (1986) Maximum-likelihood deconvolution: an optimization theory perspective. Geophysics 51:1206–1220

Hampson D, Russell B (1985) Maximum-likelihood seismic inversion. Geophysics 50(8):1380–1381

Helgesen J, Magnus I, Prosser S, Saigal G, Aamodt G, Dolberg D, Busman S (2000) Comparison of constrained sparse spike and stochastic inversion for porosity prediction at Kristin Field. Lead Edge 19(4):400–407

Kormylo JJ, Mendel JM (1983) Maximum-likelihood seismic deconvolution. IEEE Trans Geosci Remote Sens 1:72–82

Lancaster S, Whitcombe D (2000) Fast-track "colored" inversion. SEG Expanded Abstracts 19:1572–1575

Leite EP (2010) Seismic model based inversion using matlab. Matlab-Modelling, Programming and Simulations, p 389

Li Q (2001) LP sparse spike impedance inversion. Hampson-Russell Software Services Ltd, CSEG

Li LM, Speed TP (2004) Deconvolution of sparse positive spikes. J Comput Graph Statist 13(4):853–870

Loris I, Nolet G, Daubechies I, Dahlen FA (2007) Tomographic inversion using $\ell 1$-norm regularization of wavelet coefficients. Geophys J Int 170(1):359–370

Mallick S (1995) Model-based inversion of amplitude-variations-with-offset data using a genetic algorithm. Geophysics 60(4):939–954

Maurya SP, Sarkar P (2016) Comparison of post-stack seismic inversion methods: a case study from Blackfoot Field, Canada. Int J Sci Eng Res 7(8):1091–1101

Maurya SP, Singh KH (2015) Reservoir characterization using model based inversion and probabilistic neural network. Discovery 49(228):122–127

Maurya SP, Singh KH (2015b) LP and ML sparse spike inversion for reservoir characterization-a case study from Blackfoot area, Alberta, Canada. In 77th EAGE Conference and Exhibition 2015 (1):1–5

Maurya SP, Singh NP (2017) Seismic colored inversion: a fast way to estimate rock properties from the seismic data. Carbonate Reservoir Workshop, Nov. 30th–Dec. 1th, 2017, IIT Bombay, India

Maurya SP, Singh NP (2018) Application of LP and ML sparse spike inversion with probabilistic neural network to classify reservoir facies distribution—A case study from the Blackfoot Field, Canada. J Appl Geophys, Elsevier 159 (2018):511–521

Maurya SP, Singh KH (2019) Predicting porosity by multivariate regression and probabilistic neural network using model-based and colored inversion as external attributes: a quantitative comparison. J Geol Soc India 93(2):131–252

Maurya SP, Singh KH, Singh NP (2019) Qualitative and quantitative comparison of geostatistical techniques of porosity prediction from the seismic and logging data: a case study from the Blackfoot Field, Alberta, Canada. Marine Geophys Res 40(1):51–71

Mendel JM (2012) Maximum-likelihood deconvolution: a journey into model-based signal processing. Springer Science & Business Media, Berlin

Oldenburg D, Scheuer T, Levy S (1983) Recovery of the acoustic impedance from reflection seismograms. Geophysics 48(10):1318–1337

Oliveira SAM, Lupinacci WM (2013) L_1 norm inversion method for deconvolution in attenuating media. Geophys Prospect 61(4):771–777

Quijada MF (2009) Estimating elastic properties of sandstone reservoirs using well logs and seismic inversion. Doctoral dissertation, University of Calgary

Russell B (1988) Introduction to seismic inversion methods. The SEG Course Notes, Series 2

Russell B, Hampson D (1991) Comparison of post-stack seismic inversion methods. In: SEG technical program expanded abstracts, Society of Exploration Geophysicists, pp 876–878

Sacchi MD, Ulrych TJ (1995) High-resolution velocity gathers and offset space reconstruction. Geophysics 60(4):1169–1177

Sacchi MD, Ulrych TJ (1996) Estimation of the discrete fourier transform, a linear inversion approach. Geophysics 61(4):1128–1136

Stull RB (1973) Inversion rise model based on penetrative convection. J Atmos Sci 30(6):1092–1099

Swisi AA (2009) Post-and pre-stack attribute analysis and inversion of Blackfoot 3d seismic dataset. M.Sc. Thesis, University of Calgary

Velis DR (2006) Parametric sparse-spike deconvolution and the recovery of the acoustic impedance. In: SEG annual meeting. Society of Exploration Geophysicists, pp 2141–2144

Velis DR (2007) Stochastic sparse-spike deconvolution. Geophysics 73(1):R1–R9

Vestergaard PD, Mosegaard K (1991) Inversion of post-stack seismic data using simulated annealing. Geophys Prospect 39(5):613–624

Wang Y (2010) Seismic impedance inversion using l_1-norm regularization and gradient descent methods. J Inverse Ill-Posed Prob 18(7):823–838

Wang X, Shiguo Wu, Ning Xu, Zhang G (2006) Estimation of gas hydrate saturation using constrained sparse spike inversion: case study from the Northern South China. Sea Terr Atmos Ocean Sci 17(4):799–813

Waters KH, Waters KH (1987) Reflection seismology: a tool for energy resource exploration. Wiley, New York

Yilmaz O (2001) Seismic data analysis, vol 1. Society of exploration geophysicists, Tulsa, OK

Zhang R, Castagna J (2011) Seismic sparse-layer reflectivity inversion using basis pursuit decomposition. Geophysics 76:R147–R158

Zhang Q, Yang R, Meng L, Zhang T, Li P (2016) The description of reservoiring model for gas hydrate based on the sparse spike inversion. In: 7th international conference on environmental and engineering geophysics & summit forum of Chinese Academy of Engineering on Engineering Science and Technology. https://doi.org/10.2991/iceeg-16.2016.27

Chapter 4
Pre-stack Inversion

Abstract Pre-stack seismic inversion utilizes as name suggest pre-stack seismic data along with well log data to estimate P-impedance, S-impedance, density and V_P/V_S ratio away from the borehole. Pre-stack seismic inversion methods provides very high resolution subsurface images. This method provide more detail information as compared to the post-stack seismic data in which only P-impedance can be extracted but in this case one can extract simultaneously P-impedance, S-impedance, density and V_P/V_S ratio. The chapter includes details of Simultaneous inversion and elastic impedance inversion. The application of these methods to the synthetic data as well as real data are also included in this chapter.

4.1 Introduction

The pre-stack seismic inversion technique provides valuable information on rock properties such as subsurface lithology and fluid content and hence can be used for reservoir characterization. There has been enormous enhanced interest in pre-stack inversion over the past few decades or so as pre-stack inversion can be used to obtain compression and shear information from P-wave acquisition data (Goodway et al. 1997; Gray and Andersen 2000). The shear wave information is stored in the source-receiver offset (AVO) and reflection coefficients variation. To detect fluid content in reservoirs, both P-and S-wave properties of the rock are required because P-waves are sensitive to changes in pore fluid, while S-waves interact mainly with the rock matrix, which is relatively unaffected by the pore fluid. In determining the immediate hydrocarbon indices for both clastic and carbonate rocks, pre-stack inversion can be helpful and can be important for lithology and fluid discrimination (Goodway et al. 1997; Burianyk and Pickfort 2000; Gray and Andersen 2000). In past, many workers used a generalized linear reversal (GLI) method to address pre-stack inversion for rock characteristics (Tarantola 1986; Mora, 1987; Demirbag et al. 1993; Pan et al. 1994). The GLI is an iterative method that requires derivative information from an objective function and a good starting model to optimize model parameters effectively. Pre-stack inversion can also be performed using global optimization

S. P. Maurya et al., *Seismic Inversion Methods: A Practical Approach*, Springer Geophysics, https://doi.org/10.1007/978-3-030-45662-7_4

schemes such as genetic algorithm (GA) and simulated annealing (SA) (Sen and Stoffa 1991; Mallick 1995).

There are two distinct phases involved in the latest approach to estimating P-wave and S-wave rock characteristics from pre-stack seismic information (Goodway et al. 1997; Ma, 2001). The first step is to obtain ordinary reflectivities of P-and S-wave incidence through the assessment of AVO (Fatti et al. 1994). The second phase is to introduce an inversion algorithm to the reflectivity sequence derived from AVO, transform it into acoustic and shear impedances. Ma (2001) created a joint inversion method that simultaneously measures the impedances of acoustics and shears from the P-and S-wave reflectivity information derived from AVO.

Pre-stack inversion is often conducted by fitting a 3-term solution to the data, and the reliability of the results increases with increasing incident angle. The most accurate result of pre-stack inversion of P-wave seismic data is P-impedance, which can be performed on short-offset data. S-impedance estimation becomes reliable as incident angles approach 30°, whereas density evaluation (and other derived elastic constants) becomes reliable only as incident angles approach 45^0. One can invert the CMP gathers in pre-stack inversion to achieve the impedances of compression and shear-wave (Hampson et al. 2005). Pre-stack inversion converts seismic angle or offsets information into P-impedance, S-impedance, and volumes of density through seismic angle/offset gathering inclusion, well information, and fundamental stratigraphic analysis. It is similarly feasible to combine other elastic parameters such as P-impedance, V_P/V_S proportion and thickness. Typically, based on target and acquisition, two elements (P-impedance and V_P/V_S ratio) is accurate and can be used to forecast reservoir characteristics free from well control (Sherrill et al. 2008). There are many approaches available for the pre-stack inversion methods but the most famous approaches are simultaneous inversion and Elastic impedance inversion methods which are described briefly in the following sections.

4.2 Simultaneous Inversion

The simultaneous inversion method estimates simultaneously more than one parameter using pre-stack gather as input. The simultaneous inversion is based on three assumptions. The first is that the linearized approximation for reflectivity holds (Ankeny et al. 1986). The linear approximation is an approximation of a general function using a linear function. Secondly, the reflectivity section as a function of angle gather can be expressed by the Aki-Richards equations (Ma 2002). The third approximation is that there is a linear relationship between the P-impedance and S-impedance and density (Nolet 1978). Using the above three assumptions, the P-impedance, S-impedance, and density can be modeled by perturbing an initial guess model. Simmons and Backus (1996) used the following equations to invert the seismic section into impedance and density sections.

$$R_P = \frac{1}{2}\left[\frac{\Delta V_P}{V_P} + \frac{\Delta\rho}{\rho}\right] \tag{4.1}$$

$$R_S = \frac{1}{2}\left[\frac{\Delta V_S}{V_S} + \frac{\Delta\rho}{\rho}\right] \tag{4.2}$$

$$R_D = \frac{\Delta\rho}{\rho} \tag{4.3}$$

where V_P is P-wave velocity, V_S is S-wave velocity, ρ is density, R_P is reflectivity from P-wave, R_S is reflectivity from S-wave and R_D is reflectivity from density. Apart from the above discussed assumptions, Simmons and Backus (1996) also used some other assumptions which state that the reflectivity section given in Eq. 4.1 through Eq. 4.3 can be estimated from the reflectivity $R_P(\theta)$ by using the Aki-Richards approximation (Richards and Frasier 1976). The density (ρ) and P-wave velocity (V_P) are related to Gardner's equation (Gardner et al. 1974) in the following way.

$$\frac{\Delta\rho}{\rho} = \frac{1}{4}\frac{\Delta V_P}{V_P} \tag{4.4}$$

The relation between P-wave and S-wave velocity is given by the Castagna's relation (Castagna et al. 1985)

$$V_P = 1.16V_S + 1360 \tag{4.5}$$

A linearized inversion approach is used to solve for the reflectivity section. Buland and Omre (2003) have also used a similar technique which is called Bayesian linearized AVO inversion. Unlike Simmons and Backus (1996), they used three terms, $\Delta V_P/V_P$, $\Delta V_S/V_S$ and $\Delta\rho/\rho$ using the Aki-Richards approximation methods. The small reflectivity approximation is also used by some authors to relate these parameter changes to the original parameter. For changes in P-wave velocity, one can write

$$\frac{\Delta V_P}{V_P} \approx \Delta \ln V_P \tag{4.6}$$

where ln represents the natural logarithm. A similar equation can be written for S-wave velocity and density as follows.

$$\frac{\Delta V_S}{V_S} \approx \Delta \ln V_S \tag{4.7}$$

$$\frac{\Delta\rho}{\rho} \approx \Delta \ln \rho \tag{4.8}$$

Various inversion techniques for evaluating and interpreting reflection seismic data have been utilized for several decades. Some of them are based upon the Zoeppritz and Knott equation (Shuey 1985) which have been modified further; the most common incarnation is given by Aki and Richards (1980). This modification reduced 16 equations with 16 unknown parameters to a single equation with three unknown parameters of Knott-Zoeppritz under some assumptions. Now, first, derive P-impedance from the post-stack seismic inversion algorithm and then this extended to the pre-stack inversion algorithms.

4.2.1 Post-stack Inversion for P-Impedance

In this study, we extend the work of both Simmons and Backus (2003) and Buland et al. (1996) and build an approach that inverts directly for P-impedance ($Z_P = \rho V_P$), S-impedance ($Z_S = \rho V_S$), and density through an approximation similar to that of Buland and Omre (2003), using constraints similar to Simmons and Backus (1996). It is also our aim to extend this approach to post-stack impedance inversion (Hampson 1991) so that this technique can be a generalized for pre-stack and post-stack inversion methods. Therefore, we will first, review the principles of the post-stack inversion method. First, combining Eqs. 4.1 and 4.6, we will get,

$$R_{Pi} \approx \frac{1}{2}\Delta \ln Z_{pi} = \frac{1}{2}\left[\ln Z_{pi+1} - \ln Z_{pi}\right] \tag{4.9}$$

where i represents the ith the interface of the subsurface model, Z_{pi} represent p-impedance of the ith layer, Z_{pi+1} represent p-impedance at $(i+1)$th layer and R_{Pi} is P-reflectivity at ith interface. Now consider N sample points in the reflectivity series then Eq. 4.9 can be written in matrix form as follows.

$$\begin{bmatrix} R_{P1} \\ R_{P2} \\ \vdots \\ R_{PN} \end{bmatrix} = \begin{bmatrix} -1 & 1 & 0 & \cdots \\ 0 & -1 & 1 & \cdots \\ 0 & 0 & -1 & 1 \\ \vdots & \vdots & \vdots & \vdots \end{bmatrix} \begin{bmatrix} L_{P1} \\ L_{P2} \\ \vdots \\ L_{PN} \end{bmatrix} \tag{4.10}$$

where $L_{Pi} = \ln(Z_{Pi})$. Further, the convolution theorem is employed which states that the seismic trace can be generated by the convolution of the wavelet (w) with the earth's reflectivity (R) series (Ma 2002). Mathematically, it can be written as follows.

$$S_i = w_i * R_i \tag{4.11}$$

The above equation can be written in matrix form as follows.

$$\begin{bmatrix} S_1 \\ S_2 \\ \vdots \\ S_N \end{bmatrix} \begin{bmatrix} w_1 & 0 & 0 & \cdots \\ w_2 & w_1 & 0 & \cdots \\ w_3 & w_2 & w_1 & 0 \\ \vdots & \vdots & \vdots & \vdots \end{bmatrix} \begin{bmatrix} L_{P1} \\ L_{P2} \\ \vdots \\ L_{PN} \end{bmatrix} \tag{4.12}$$

where S_i represent ith sample of the seismic trace and w_j represent the jth term of the extracted seismic wavelet. Combining Eqs. 4.11 and 4.12 gives the forward model which relates the seismic trace to the logarithm of P-impedance as follows.

$$T = \frac{1}{2} W D L_P \tag{4.13}$$

where W is the wavelet matrix given in Eq. (4.12) and D is the derivative matrix given in Eq. (4.11). If Eq. (4.13) is inverted using standard matrix inversion methods to give an estimate of L_P from knowledge of seismic trace (S) and wavelet (W), there are two problems. First, matrix inversion is costly and second, it is potentially unstable (Hampson et al. 2005). Now, the post-stack inversion is extended to the pre-stack inversion as follows.

4.2.2 Extension to Pre-stack Inversion

Now, in this section, the derivation is extended for the pre-stack inversion methods. One can redefine the Aki-Richards equation (Fatti et al. 1994) as follows.

$$R_{PP}(\theta) = c_1 R_P + c_2 R_S + c_3 R_D \tag{4.14}$$

where $c_1 = 1 + \tan^2\theta$, $c_2 = -8\gamma^2 \tan^2\theta$, $c_3 = -0.5\tan^2\theta + 2\gamma^2 \sin^2\theta$ and $\gamma = V_S / V_P$ and the three reflectivity terms are as given by Eqs. 4.1, 4.2 and 4.3 (Hampson et al. 2005).

For a given trace $S(\theta)$ extending the zero offset (or angle) trace given in Eq. 4.12 by combining it with Eq. 4.14, one can get the following equation.

$$S(\theta) = \frac{1}{2} c_1 W(\theta) D L_P + \frac{1}{2} c_2 W(\theta) D L_S + c_3 W(\theta) D L_D \tag{4.15}$$

where $L_S = \ln(Z_S)$ and $L_D = \ln(\rho)$. Equation 4.15 could be used for inversion, except that it ignores the relationship between L_P with L_S and L_D. Because we are dealing with impedance and have taken logarithms, therefore our relationships are different than those given by Simmons and Backus (1996) and are given as follows.

$$\ln(Z_S) = k \ \ln(Z_P) + k_C + \Delta L_S \tag{4.16}$$

$$\ln(Z_D) = m \ \ln(Z_P) + m_C + \Delta L_D \tag{4.17}$$

ΔL_S and ΔL_D can be estimated from the trend of the cross plot of $\ln(\rho)$ versus $\ln(Z_P)$ and $\ln(Z_P)$ versus $\ln(Z_S)$ respectively (Fig. 4.1).

Now, combining Eq. 4.15 through 4.17, one get the following equation.

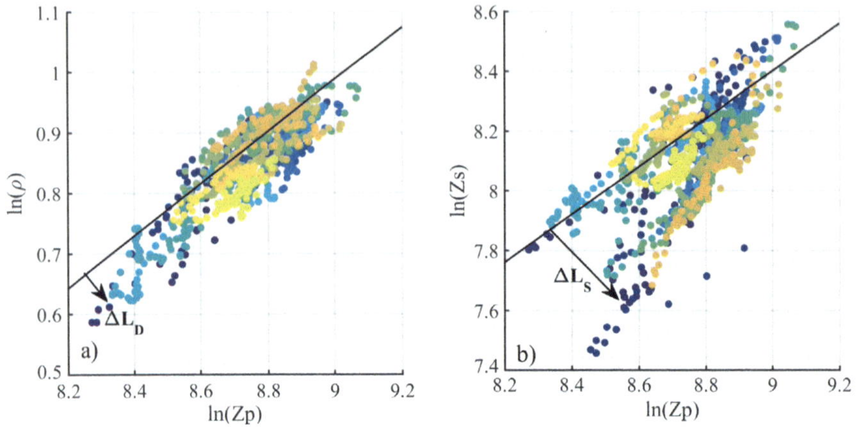

Fig. 4.1 Crossplot of **a** $\ln(Z_P)$ versus $\ln(\rho)$ and **b** $\ln(Z_P)$ versus $\ln(Z_S)$ wherein both cases a best fit straight line is added. The deviation away from this straight line ΔL_S and ΔL_D are the desired fluid anomalies

$$T(\theta) = C_1 W(\theta) * DL_P + C_2 W(\theta) * DL_S + C_3 W(\theta) * DL_D \tag{4.18}$$

where $C_1 = \frac{1}{2}c_1 + \frac{1}{2}kc_2 + mc_3$ and $C_2 = \frac{1}{2}c_2$. Equation 4.18 can be implemented in matrix form as follows.

$$\begin{bmatrix} S(\theta_1) \\ S(\theta_2) \\ \vdots \\ S(\theta_N) \end{bmatrix} \begin{bmatrix} C_1(\theta_1)W(\theta_1)D & C_2(\theta_1)W(\theta_1)D & C_3(\theta_1)W(\theta_1)D \\ C_1(\theta_2)W(\theta_2)D & C_2(\theta_2)W(\theta_2)D & C_3(\theta_3)W(\theta_3)D \\ \vdots & \vdots & \vdots \\ C_1(\theta_N)W(\theta_N)D & C_2(\theta_N)W(\theta_N)D & C_3(\theta_N)W(\theta_N)D \end{bmatrix} \tag{4.19}$$

If Eq. 4.19 is solved by matrix inversion methods, then we again have the same problem that the low-frequency component cannot be resolved properly. So a practical approach is to initialize the solution to $[L_P \Delta L_S \Delta L_D] = [\ln Z_{P0} 00]$, where Z_{P0} is the initial impedance model (Larson 1999). Equation 4.19 is used for the inversion of pre-stack seismic data. The practical of the inversion can be written as follows.

i. The following information from the seismic section is available for the input.
   - One set of N angle traces;
   - One set of N wavelets for each angle;
   - Initial model for Z_P.
ii. Next, calculate the k and m coefficient using well-log cross plot as discussed above.
iii. Generate the initial guess model by interpolating well log property in the seismic section.

$$[L_P \Delta L_S \Delta L_D]^T = [\ln Z_{P0} 0 0]^T \tag{4.20}$$

iv. Apply the pre-stack inversion technique discussed above.
v. Finally, calculate Z_P, Z_S and density as follows.

$$Z_P = \exp(L_P) \tag{4.21}$$

$$Z_S = \exp(kL_P + k_c + \Delta L_S) \tag{4.22}$$

$$\rho = \exp(mL_P + m_c + \Delta L_D) \tag{4.23}$$

The initial guess model representing the initial model of P-impedance while ΔL_S and ΔL_D are initialized with zero values in this iteration (Maurya and Singh 2015, 2018). Further, this impedance can be transformed into lame parameters which are discussed in detail in the following section.

4.2.3 Lambda-Mu-Rho (LMR) Transform

The LMR method was originally proposed by Goodway et al. (1997). The lambda rho ($\lambda\rho$) and mu rho ($\mu\rho$) parameters are more sensitive to the fluids and saturation compare to the velocity and density and therefore analysis in the lambda-mu-rho domain would be more beneficial (Russell 1988). The LMR transform uses Z_P and Z_S from the simultaneous inversion and transform them into lambda rho and mu rho sections. The LMR parameters can be derived as follows.

The P-wave (V_P) and S-wave (V_S) velocity can be written in terms of lame parameters as follows.

$$V_P = \sqrt{\frac{\lambda + 2\mu}{\rho}} \tag{4.24}$$

And

$$V_S = \sqrt{\frac{\mu}{\rho}} \tag{4.25}$$

Where λ and μ are lame parameters and ρ is density. Thereafter, Eq. 4.25 can be rewritten as follows.

$$Z_S^2 = (\rho V_S)^2 = \mu\rho \tag{4.26}$$

Further, Eq. 4.24 can be written as follows.

$$Z_P^2 = (\rho V_P)^2 = (\lambda + 2\mu)\rho \tag{4.27}$$

Thereafter, combining Eqs. 4.26 and 4.27, we have the following relation.

$$\lambda\rho = Z_P^2 - 2Z_S^2 \tag{4.28}$$

The lame parameters are derived by extracting the P-wave and S-wave reflectivities from pre-stack seismic inversion and inverting them to P-wave and S-wave impedances. Equations 4.26 and 4.28 are used in the analysis of LMR transform.

The LMR analysis is very crucial to identify gas sands (Goodway et al. 1997). This can be done from the separation of in responses of both the $\lambda\rho$ and $\mu\rho$ sections to gas sands versus shales. In some other cases, the LMR transform is used to separate lithologies at an even finer scale so as to identify wet sands from shales. The LMR cross plot is also used to discriminate subsurface lithology (Paul et al. 2001). Separately neither lambda-rho ($\lambda\rho$) nor mu-rho ($\mu\rho$) section is a good indicator of subsurface lithology but together they can be used to reveal a great deal about lithology. A flowchart of pre-stack simultaneous inversion and LMR transform has been presented in Fig. 4.2.

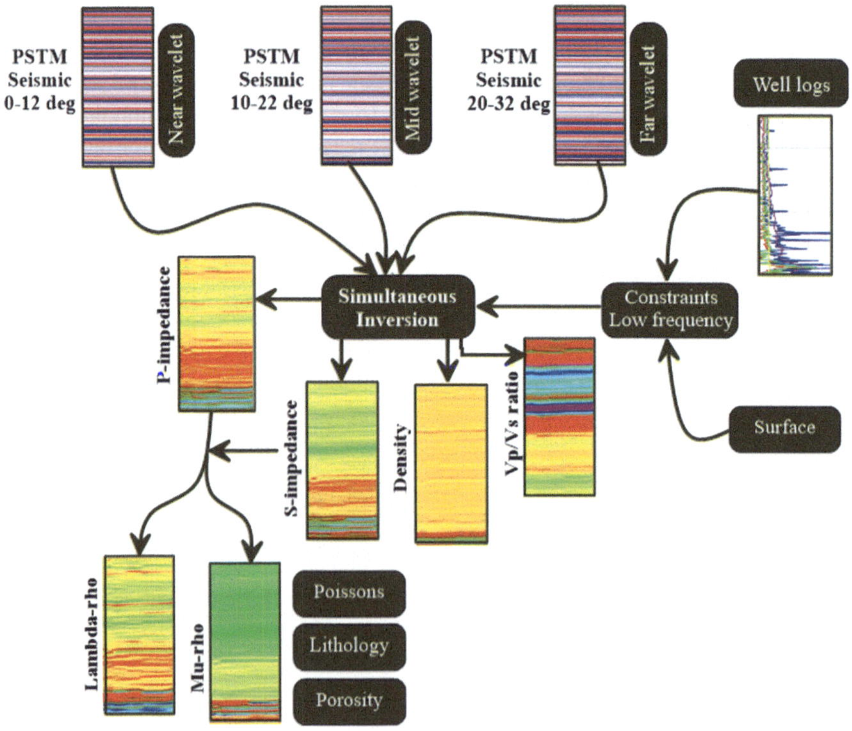

Fig. 4.2 Flowchart of pre-stack simultaneous inversion methods

Now we are in a position to implement the above methods to the practical data sets. Two examples are presented, the first for the synthetic data sets where the true output is known and hence one can verify their results and second is for the example from the real data sets from the Penobscot field, Nova Scotia, Canada.

4.2.4 Application of Simultaneous Inversion

The first example for the pre-stack simultaneous inversion is from the synthetically generated data sets. For the generation of synthetic seismograms, the parameters given in Table 4.1 are used as input and forward modeling procedures are performed to generate synthetic seismograms with angle-dependent. Figure 4.3 demonstrates a synthetic example for pre-stack simultaneous inversion methods. The first track of Fig. 4.3 depicts subsurface lithology model used to generate synthetics, track 2 depicts reflectivity series generated from the velocity and density combination given in geological model, track 3 shows modeled synthetic seismic trace generated using forward modeling for nonzero offset case (Eq. 1.10), track 4, track 5, track 6 and track 7 demonstrate comparison of inverted Z_P, Z_S, ρ and V_P/V_S ratio with their original values, respectively. From the figure, one can notice that the inverted curves are following the trend of the original curve very well. The estimated correlation is 0.99 and RMS error is 0.023 for all four inverted curves which also indicate good results and hence good performance of the simultaneous inversion methods.

The other example shown here from the real datasets of the Penobscot field, Nova Scotia, Canada. The Penobscot survey area lies in the Scotian shelf in the Offshore Nova Scotia, Canada. Several minor oil and gas fields are located in the Late Jurassic to Cretaceous intervals. The seismic data used for pre-stack simultaneous inversion need to be conditioned to remove as many undesirable effects. Three major undesirable effects commonly removed by data conditioning are random noise, NMO wavelet stretch, and non-flat reflections. The conditioning of gather is briefly explained in Chap. 3 and hence here we have used trim gather as input for simultaneous inversion.

In a pre-stack simultaneous inversion, invert the pre-stack CMP gathers to obtain the compressional wave impedance (Z_P), shear-wave impedances (Z_S), density (ρ) and V_P/V_S ratio. These parameters are very important to interpret subsurface features. Like other inversion techniques discussed in Chap. 3, except colored inversion, building of an initial model is required for pre-stack inversion. The model is built by using the P-wave, S-wave velocity and density logs at well locations. From these logs, the P- and S-impedance logs are estimated ($Z = V_P\rho$), and interpolated between the wells to build the models by using horizons as guides. The models are filtered by using a 12-Hz low-pass filter to remove the high-frequency component and well-log heterogeneity at the seismic frequencies band (Shu-jin 2007). The details of model building are shown in Chap. 3.

Pre-stack simultaneous inversion relies on the background relationship between $\ln(Z_P)$; $\ln(Z_S)$, and $\ln(\rho)$ at the well locations, from which the coefficients

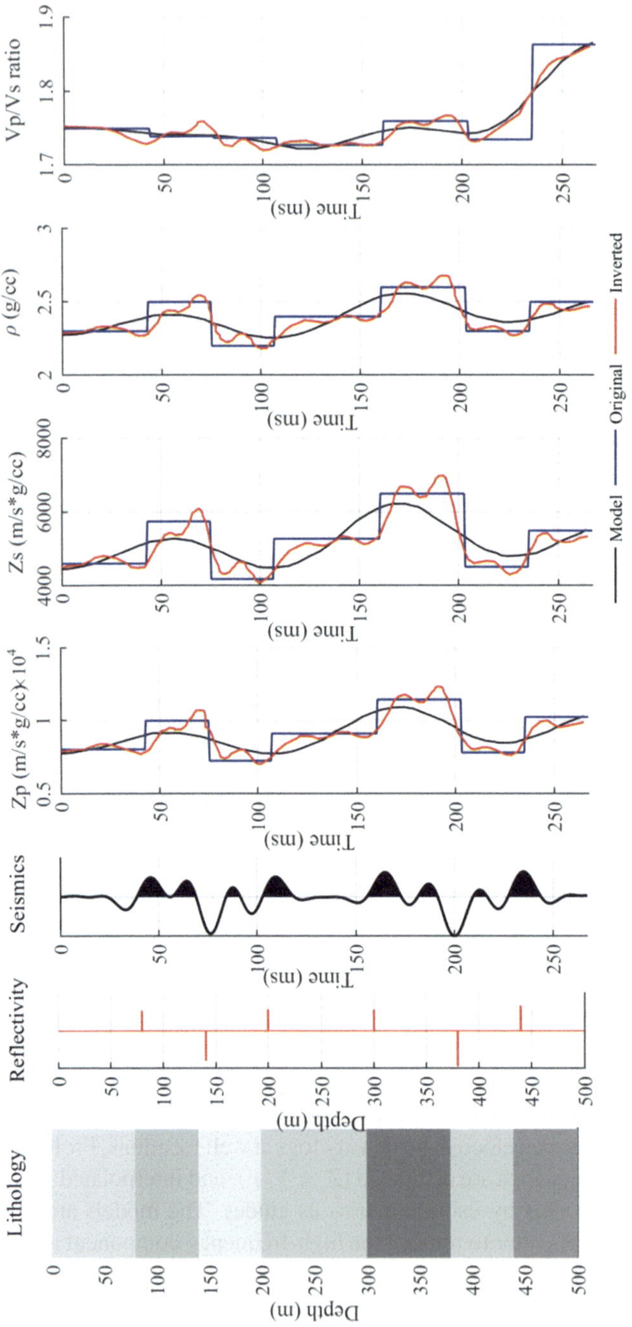

Fig. 4.3 Synthetic example of pre-stack simultaneous inversion methods

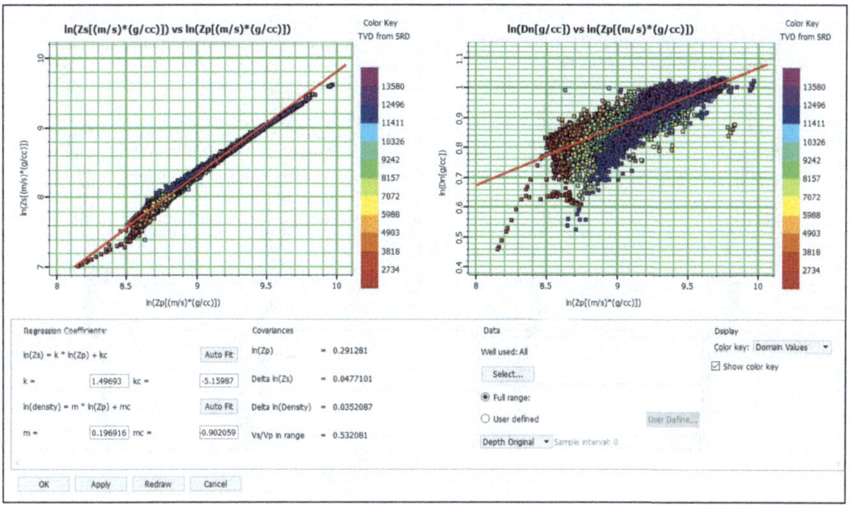

Fig. 4.4 Crossplot of $\ln(Z_P)$ versus $\ln(Z_S)$ (left) and $\ln(Z_P)$ versus $\ln(\rho)$ (right). The measured parameters from these cross plots are shown bottom of the figures

(k; k_c; m and m_c) are calculated. Deviations of these values from the background ΔL_S and ΔL_D are calculated from the inversion itself, and therefore they are initialized equal to zero in the initial model. When the coefficients (k; k_c; m and m_c)) are calculated, they are utilized to determine the final inversion (Shu-jin 2007). These values are given in Fig. 4.4.

After determining the coefficients of simultaneous inversion, the pre-stack inversion is performed in two stages. In first stage, we apply the inversion at the well locations for one composite trace to test the parameters and scale the seismic data. Figure 4.5 compares the inversion result of the P-wave impedance (Z_P), S-wave impedance (Z_S), density (ρ) and the velocity ratio V_P/V_S with the corresponding parameters at well (Well-L-30). From the result, one can notice that the inverted curves are matching with the original curves very nicely for all parameters. The average correlation between them is 0.94 and RMS error is 1100 m/s g/cc.

Further, for the quality check of inversion results point by point, a cross plot of inverted and original curves are generated and displayed in Fig. 4.6. The figure has four subplots corresponding to compressional wave impedance (Z_P), shear-wave impedances (Z_S), density (ρ) and V_P/V_S ratio. The distribution of scatter points from the best fit straight line quantifies the quality of inversion results. It can be noticed that most data points lie very close to the best fit line which indicates good quality of inversion results.

The analysis shows that the simultaneous inversion works well for composite trace. In the second step entire seismic volume is inverted for the petrophysical volumes. Figure 4.7 shows a cross-section of inverted acoustic impedance. Inversion of seismic amplitudes generated the color-coded panel, with low acoustic impedance in green and yellow, and high acoustic impedance in pink and blue. The acoustic impedance

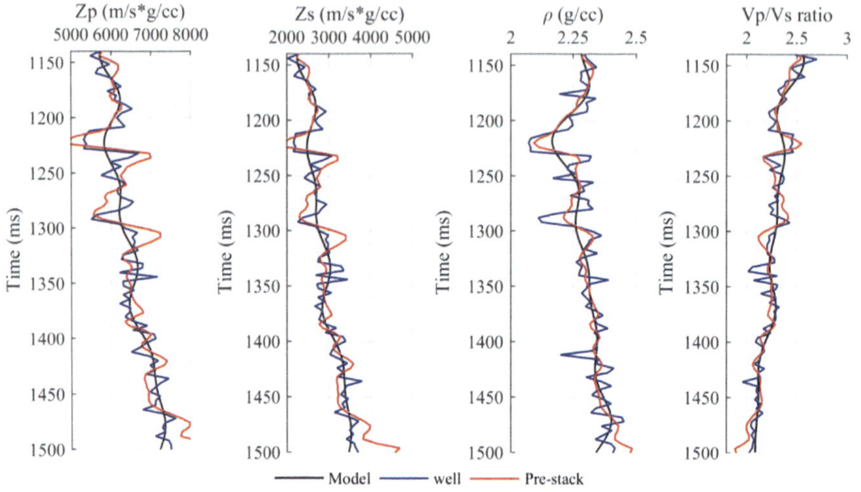

Fig. 4.5 Inversion analysis results for composite trace near to well location

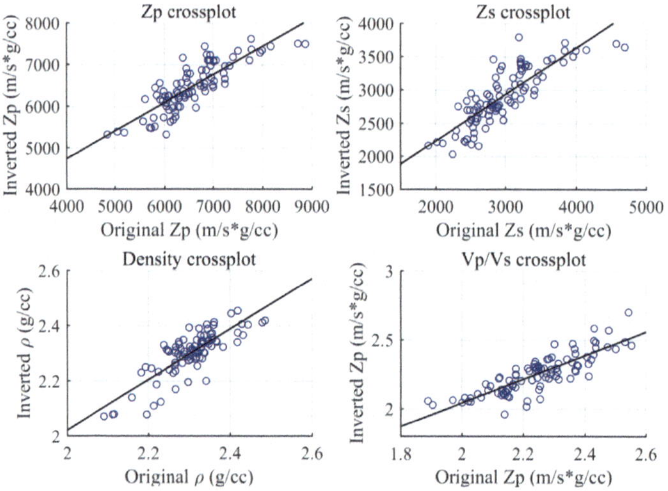

Fig. 4.6 Crossplot of inverted curves versus original curves

of the area varies from 2000 m/s*g/cc to 8000 m/s g/cc. The P-impedance section (inline 1165) shows smooth variation along with the vertical as well as horizontal direction. From results, one can also notice that the area has no major anomaly zone. The shear impedance section (inline 1165) is shown in Fig. 4.8, the density section (inline 1165) is shown in Fig. 4.9 and the velocity ratio section (inline 1165) is shown in Fig. 4.10. The shear impedance varies from 100 m/s*g/cc to 4000 m/s g/cc and the density varies from 2 to 2.5 g/cc. These figures also show smooth variation along

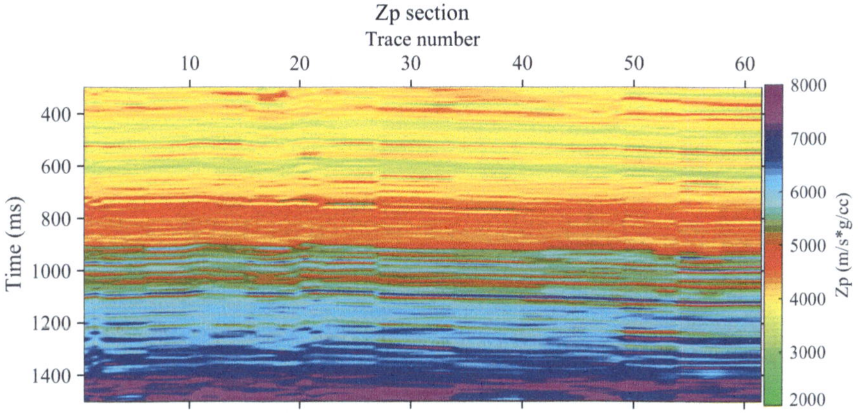

Fig. 4.7 Cross-section of inverted P-impedance at inline 1164

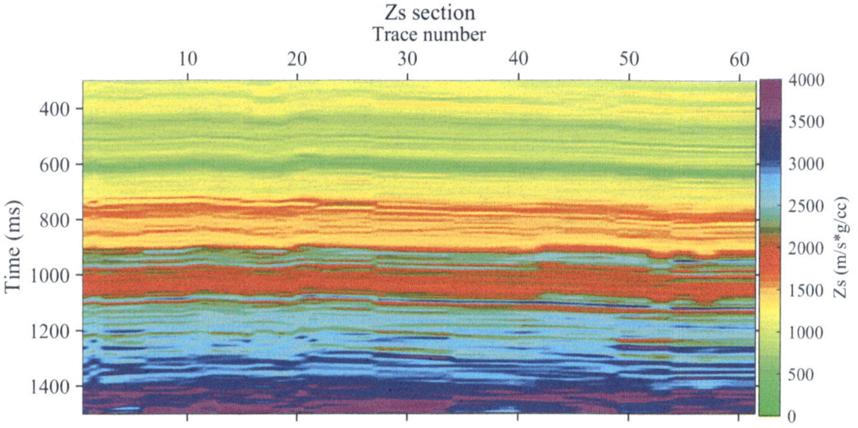

Fig. 4.8 Cross-section of inverted S-impedance at inline 1165

with the vertical and horizontal and hence no indication of any major anomaly zone. If any data contains a reservoir, then low impedance and density anomaly should appear.

Further, one also uses these sections to estimate the relationship between P-impedance, S-impedance, and density. Cross plot among these parameters is provided in Fig. 4.11. This cross plot also provide a derived relationship. These relationships are better to estimate one of these parameters if one knows the other as compared with the relationship derived from well logs which are valid only at particular locations and not away from the well location.

In some situations, the impedance and velocity used to differentiate fluid and rock properties and hence need other properties to discriminate subsurface lithology. The lame parameters are found more useful in the situation discussed above. These lame

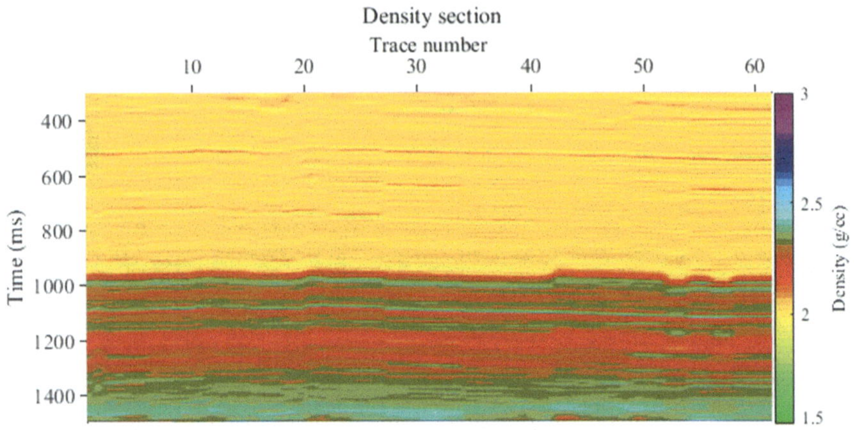

Fig. 4.9 Cross-section of inverted density at inline 1165

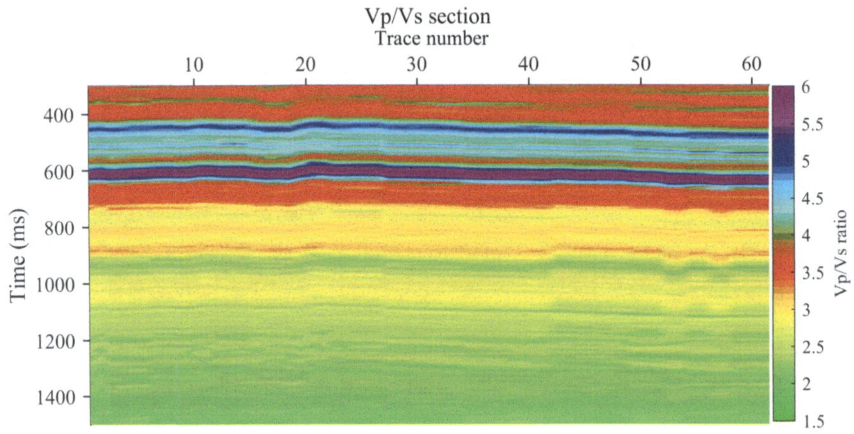

Fig. 4.10 Cross-section of inverted V_P/V_S at inline 1165

parameters are more sensitive and vary more rapidly than impedance and velocity and hence indicate more clear fluid discrimination.

The Z_P and Z_S is transformed into lame parameters using LMR transform approach discussed in the above section. The LMR feature transforms S- and P-impedance volumes directly into Lambda-Rho and Mu-Rho volumes. This simple yet powerful transform allows getting more physically meaningful information out of inversion results. The workflow of the LMR transform is as follows.

- In the first step; calculate R_P and R_S reflectivities from pre-stack data.
- Thereafter, use the simultaneous inversion method to estimate Z_P and Z_S volumes from the AVO and inversion functions.

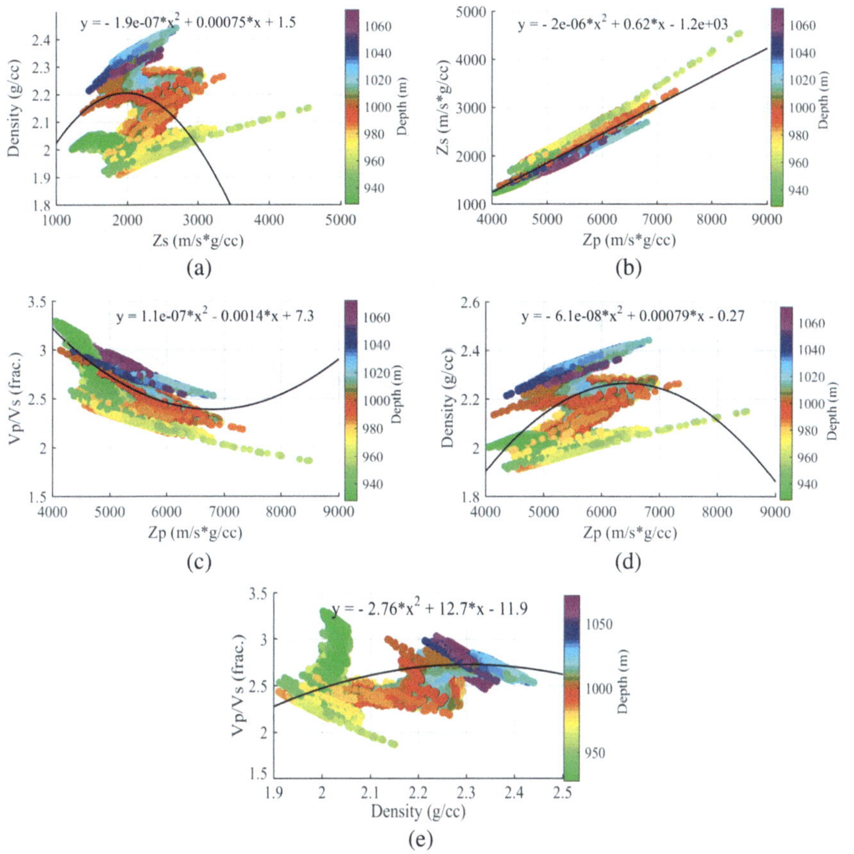

Fig. 4.11 Crossplot of **a** density versus S-impedance, **b** P-impedance versus S-impedance, **c** P-impedance versus V_P/V_S ratio, **d** Density versus P-impedance, and **e** density versus V_P/V_S ratio

- Further, impedances volumes are transformed into $\lambda\rho$ and $\mu\rho$ using Eqs. 4.26 and 4.28 respectively. Note that we cannot decouple density from the other terms.
- Draw cross plot of $\lambda\rho$ versus $\mu\rho$ to minimize the effect of density. From this cross plot, one can classify hydrocarbon sands from wet sands.

Figure 4.8 depicts variation of modeled $\lambda\rho$ cross-section (inline 1165) whereas $\mu\rho$ cross-section (inline 1165) is shown in Fig. 4.9. The modeled section shows very high resolution as compared with input seismic data and describes layer properties. One can notice that the $\lambda\rho$ varies from 5 to 30GPa g/cc whereas the $\mu\rho$ varies from 0 to 10GPa g/cc. The analysis of modeled lame parameters volume demonstrates smooth variation along vertical as well as horizontal and hence confirms that the area does not have any major anomaly zone.

Further, a cross plot of modeled $\lambda\rho$ and $\mu\rho$ is generated and is shown in Fig. 4.10. There are no data points near to lambda-rho and mu-rho axis (no low values) which

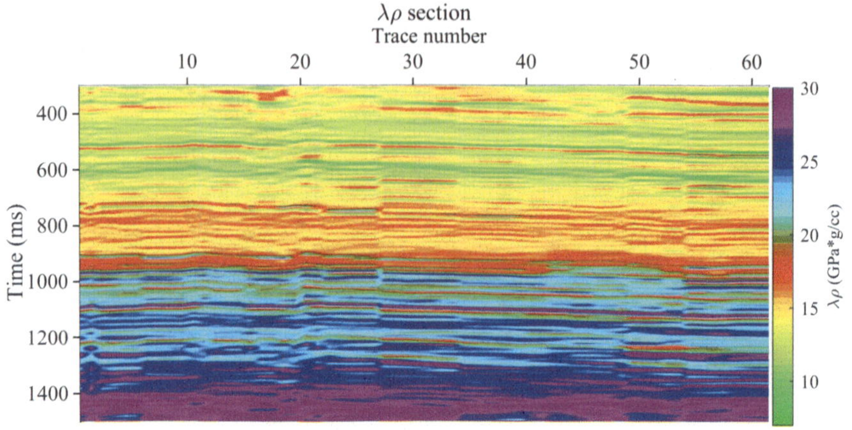

Fig. 4.12 Cross-section of inverted lambda-rho ($\lambda\rho$) at inline 1165

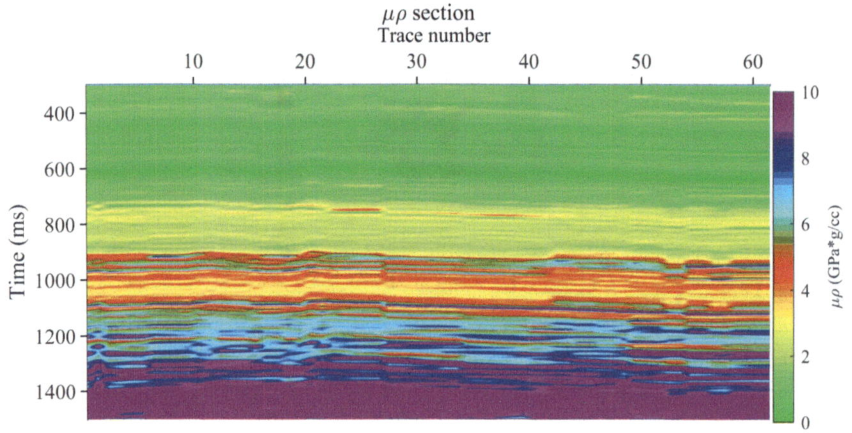

Fig. 4.13 Cross-section of inverted mu-rho ($\mu\rho$) at inline 1165

shows that there is no major reservoir in the study area. Lambda-rho and mu-rho parameters are very important in the gas reservoir because these parameters are more sensitive compared to the impedances (Figs. 4.12, 4.13 and 4.14).

4.3 Elastic Impedance Inversion

The Elastic impedance (EI) inversion is another approach of pre-stack inversion methods which represents a generalization of the acoustic impedance inversion of offset dependent angle gather. Because *P*/*S* mode conversions are significant at oblique

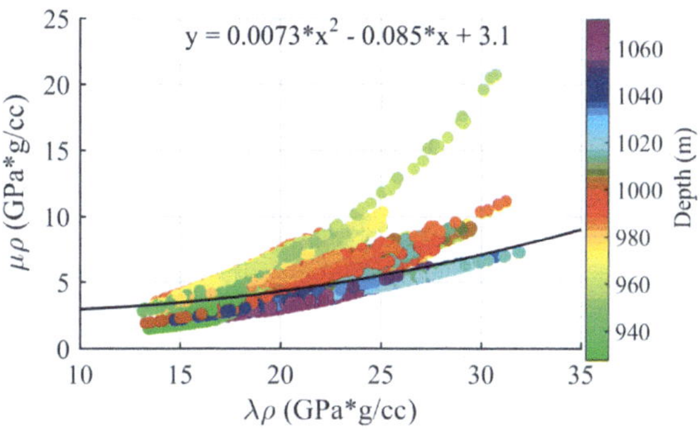

Fig. 4.14 Crossplot of inverted lambda-rho versus mu-rho shows the relationship between them

incidence angle hence EI inversion gives a more detailed picture of the subsurface (Whitcombe 2002). The EI inversion can be derived from the small-contrast Aki and Richards's equation (Eq. 4.29). Connolly (1999) was the first to use range-limited stack section to invert for elastic impedance. The reflection coefficient $R(\theta)$ can be calculated by using the Zeopritz equation for P-wave reflectivity for an angle θ as follows.

$$R(\theta) = A + B\sin^2\theta + C\sin^2\theta\tan^2\theta \tag{4.29}$$

where $A = \frac{1}{2}\left(\frac{\Delta V_P}{V_P}\right)$, $B = \frac{\Delta V_P}{V_P} - 4\frac{V_S^2}{V_P^2}\frac{\Delta V_S}{V_S} - 2\frac{V_S^2}{V_P^2}\frac{\Delta\rho}{\rho}$ and $C = \frac{1}{2}\frac{\Delta V_P}{V_P}$ and

$$V_P = \frac{V_P(t_i) + V_P(t_{i-1})}{2},\ \Delta V_P = V_P(t_i) + V_P(t_{i-1}),$$
$$\frac{V_S^2}{V_P^2} = \left(\frac{V_S^2(t_i)}{V_P^2(t_i)} - \frac{V_S^2(t_{i-1})}{V_P^2(t_{i-1})}\right)/2$$

and similarly for the other variables. In Eq. 4.29, V_P represents P-wave velocity, V_S represents S-wave velocity, ρ is density and t_i is the time at sample i.

Further, one requires a function f(t) which has properties analogous to acoustic impedance, such that reflectivity can be derived from the formula given for any incidence angle θ (Tarantola 1986). Mathematically, it can be written as follows.

$$R(\theta) = \frac{f(t_i) - f(t_{i-1})}{f(t_i) + f(t_{i-1})} \tag{4.30}$$

Now, this function called elastic impedance (EI), and then for small to moderate changes in impedance, the above expression can be written as

$$R(\theta) \approx \frac{1}{2}\frac{\Delta EI}{EI} \approx \frac{1}{2}\Delta\ln(EI) \tag{4.31}$$

And so, Eq. 4.29 can be rewritten as follows.

$$\frac{1}{2}\Delta\ln(EI) = \frac{1}{2}\left(\frac{\Delta V_P}{V_P} + \frac{\Delta\rho}{\rho}\right) + \left(\frac{\Delta V_P}{2V_P} - 4\frac{V_S^2}{V_P^2}\frac{\Delta V_S}{V_S} - 2\frac{V_S^2}{V_P^2}\frac{\Delta\rho}{\rho}\right)\sin^2\theta$$
$$+\frac{1}{2}\frac{\Delta V_P}{V_P}\sin^2\theta\tan^2\theta \quad (4.32)$$

Let $K = \frac{V_S^2}{V_P^2}$, then Eq. 4.32 can be rewritten as follows.

$$\frac{1}{2}\Delta\ln(EI) = \frac{\Delta V_P}{2V_P} + \frac{\Delta\rho}{2\rho} + \frac{\Delta V_P}{2V_P}\sin^2\theta - 4K\sin^2\theta\frac{\Delta V_S}{V_S}$$
$$-2K\frac{\Delta\rho}{\rho}\sin^2\theta + \frac{1}{2}\frac{\Delta V_P}{V}\sin^2\theta\tan^2\theta \quad (4.33)$$

In the rearranging term, we will get the following.

$$\frac{1}{2}\Delta\ln(EI) = \frac{1}{2}\left[\frac{\Delta V_P}{V_P}(1+\sin^2\theta) + \frac{\Delta\rho}{\rho}(1-4K\sin^2\theta)\right.$$
$$\left.-\frac{\Delta V_S}{V_S}8K\sin^2\theta + \frac{\Delta V_P}{V_P}\sin^2\theta\tan^2\theta\right] \quad (4.34)$$

Using $\sin^2\theta\tan^2\theta = \tan^2\theta - \sin^2$, so Eq. (4.6) becomes as follows.

$$\frac{1}{2}\Delta\ln(EI) = \frac{1}{2}\left[\frac{\Delta V_P}{V_P}(1+\tan^2\theta) - \frac{\Delta V_S}{V_S}8K\sin^2\theta + \frac{\Delta\rho}{\rho}(1-4K\sin^2\theta)\right] \quad (4.35)$$

We used only the first two terms of Eq. (4.29), then the above and following expressions differ only by changing the $\tan^2\theta$ to $\sin^2\theta$ (Whitcombe 2002). We substitute again $\Delta\ln x$ for $\Delta x/x$;

$$\Delta\ln(EI) = (1+\tan^2\theta)\Delta\ln(V_P) - 8K\sin^2\theta\Delta\ln(V_S) + (1-4K\sin^2\theta)\Delta\ln(\rho) \quad (4.36)$$

Now if we make K a constant we can take all terms inside the Δs;

$$\Delta n(EI) = \Delta\ln\left(V_P^{(1+\tan^2\theta)}\right) - \Delta\ln\left(V_S^{(8K\sin^2\theta)}\right) + \Delta\ln\left(\rho^{(1-8K\sin^2\theta)}\right) \quad (4.37)$$

$$\Delta\ln(EI) = \Delta\ln\left(V_P^{(1+\tan^2\theta)}V_S^{(8K\sin^2\theta)}\rho^{(1-8K\sin^2\theta)}\right) \quad (4.38)$$

Finally, we integrate and exponentiate (i.e., remove the differential and logarithmic terms on both sides), setting the integration constant to zero. We have the following equation.

$$EI = V_P^{(1+\tan^2\theta)} V_S^{(8K\sin^2\theta)} \rho^{(1-8K\sin^2\theta)} \tag{4.39}$$

Equation (4.39) is the final output which is used to estimate elastic impedance if one knows the P-wave velocity, S-wave velocity, density and angle θ (Connolly 1999; Lu and McMechan 2004; Mallick 2001; Maurya 2019).

4.3.1 Application of Elastic Impedance Inversion

Real data from the Penobscot field, Canada is utilized to demonstrate the capability of EII to estimate subsurface elastic properties. Pre-stack seismic data from inline 1161–1200 and cross-line 1001–1481 along with well L-30 is used as input. Initially, data preparation is performed which includes a partial stack, Horizon picking, depth to time conversion, wavelet extraction and generation of the low-frequency initial model. Figure 4.15 shows near angle stack (left) and far angle stack (right) which is used as input for elastic impedance inversion methods. Figure 4.15 depicts that the

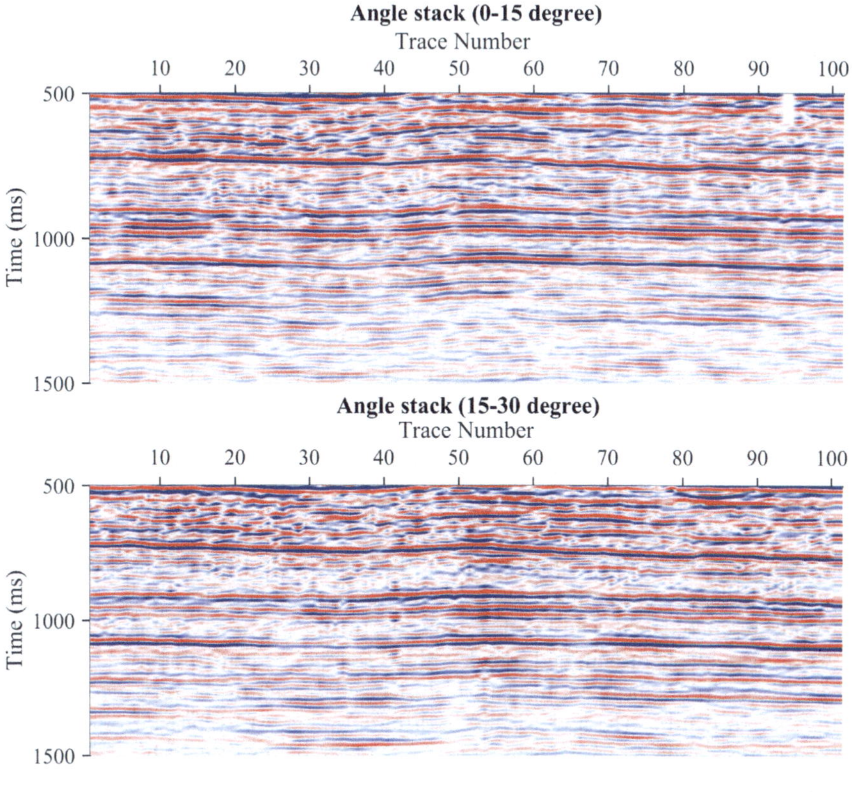

Fig. 4.15 Cross-section of near angle stack (top) and far angle stack (bottom) at inline 1165

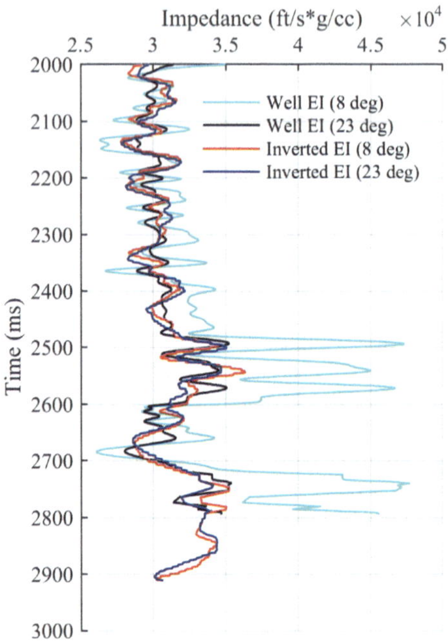

Fig. 4.16 Comparison of inverted and original elastic impedance (EI) for the composite trace

signal to noise ratio is increased due to stacking which is the usual case but it is also noticed that the resolution of the far stack section increases higher as compared to the near stack gather (highlighted by arrow). The near stack is considered by a 0°–15° incidence angle while the far stack is considered for 15°–30°. After preparing data, elastic impedance inversion is applied to data in two steps, first, one composite trace near to well a location is extracted from near as well as far angle stack gather and inversion is performed to this trace and results are compared with well log EI.

Figure 4.16 depicts a comparison of inverted elastic impedance from the near stack as well as a far stack with well log impedance. It is noticed from Fig. 4.16 that the inverted EI for the near stack as well as far stack are agreeing nicely with the well log EI. The estimated average correlation is very high (0.94) and error is 0.13 which depict the good performance of the algorithm. For the quality check, the cross plot is generated between original and inverted elastic impedance for near (left) as well as far angle stack (right) data and displayed in Fig. 4.17. From Fig. 4.17, it is noticed that the maximum scatter points lie near to the best fit line for both cases (near and far angle stack) and depict the good performance of the algorithm.

As elastic impedance is an extension of acoustic impedance (AI) hence a comparison of EI from near and far angle stack with AI is generated and plotted in Fig. 4.18. From Fig. 4.18, it is noticed that EI is following the trend of AI very nicely. Generally, both quantities (EI and AI) are in different range as EI varies from 4000 to 8000 m/s g/cc while AI varies from 5000 to 12,000 m/s g/cc hence shows deviation with each other particularly at larger impedance zone (Fig. 4.18). A cross plot of near stack elastic impedance with far stack elastic impedance is generated with a color

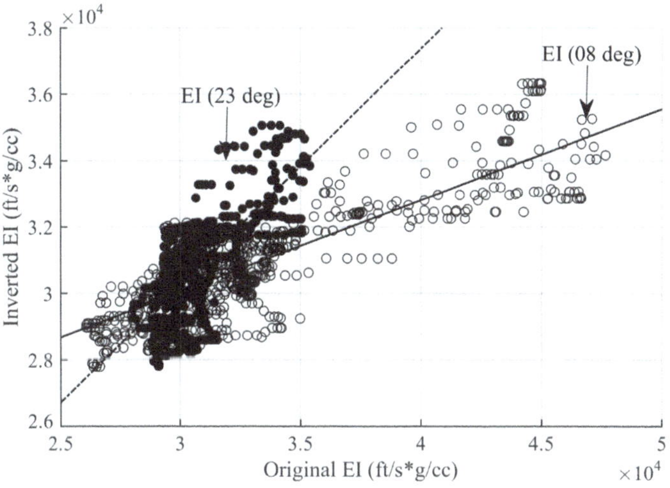

Fig. 4.17 Cross-plot of inverted and original elastic impedance for the near-angle stack and far-angle stack

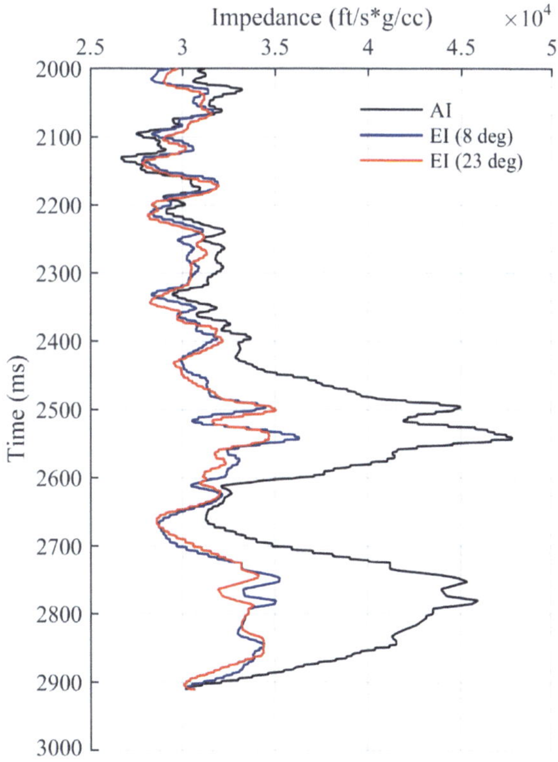

Fig. 4.18 Comparison of elastic impedance (near and far) with acoustic impedance for composite trace near a well location

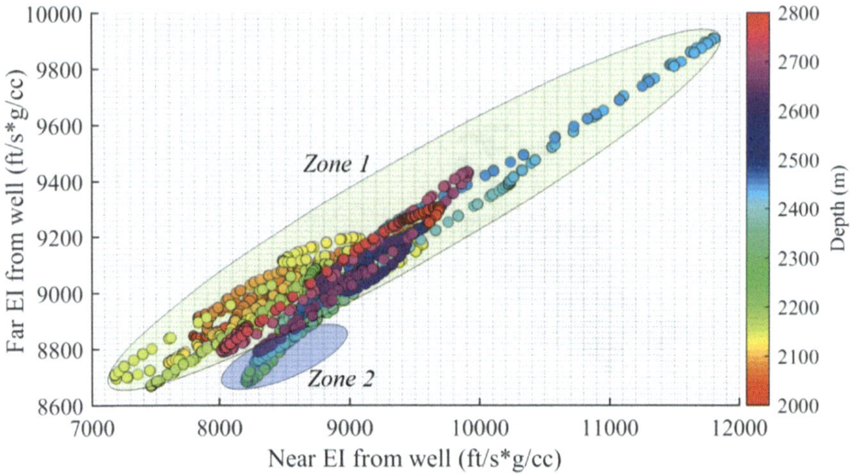

Fig. 4.19 Cross-plot of inverted near stack elastic impedance versus far angle stack impedance. The two different zones are highlighted

bar of acoustic impedance and is shown in Fig. 4.19. From Fig. 4.19, it is noticed that both EI (near and far angle stack) vary in a similar way. The whole cross plot is divided into two-zone, zone 1 and zone 2. Zone 1 corresponds to maximum data points which also follow a trend. On the other hand, zone 2 contains data points that fall away from the trend line and indicates an anomalous zone. These small cluster points indicate the presence of gas accumulation in the subsurface. The gas accumulation is very minor.

Thereafter, the entire seismic section is inverted for elastic impedance in the second step and cross-section (inline 1165) is shown in Fig. 4.9. The upper side of the figure shows near angle stacks elastic impedance while far angle stack elastic impedances are shown bottom of Fig. 4.9. Both sections show the variation of elastic impedance vertically as well as horizontally in the subsurface. The improved reflector resolution has been noticed from both the figures but it is higher in the far stack elastic impedance section as compared with near stack elastic impedance section comparatively.

A cross plot is always used to see any anomalous zone present in the area and hence, near stack and far stack elastic impedance is generated for entire datasets and shown in Fig. 4.10. The whole cross plot is divided into two-zone, zone 1 and zone 2. Zone 1 corresponds to maximum data points which also follow a trend. On the other hand, zone 2 contains data points that fall away from the trend line and indicates an anomalous zone. These small cluster points indicate the presence of gas formation in the subsurface. Although, zone 2 shows very minor accumulation of gas which cannot be detected from the seismic section as well as inverted impedance sections. The location of this minor accumulation of gas is shown in the seismic section and presented in Fig. 4.15. This is standard way to present anomalous zone in the input

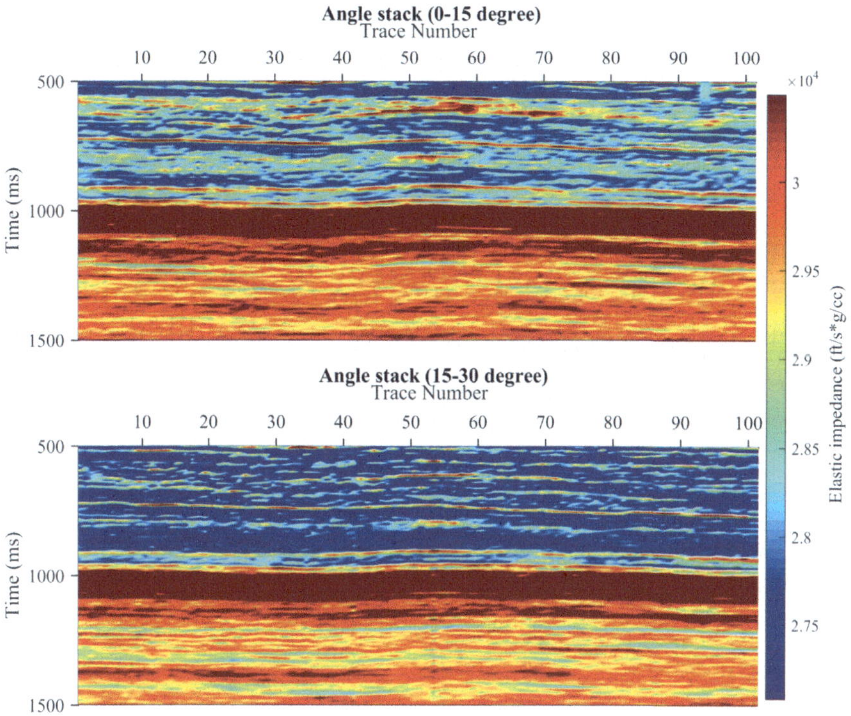

Fig. 4.20 Cross-section of inverted elastic impedance from near angle stack (top) and far angle stack (bottom) at inline 1165

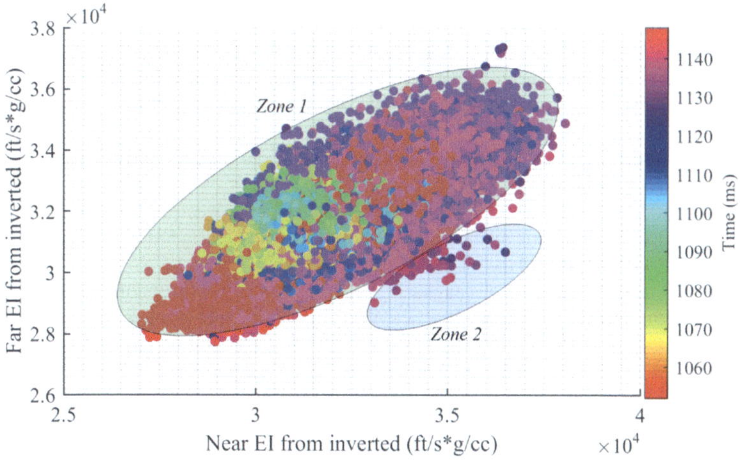

Fig. 4.21 Crossplot of inverted elastic impedance from near angle stack versus far angle stack data. Two different characteristics are highlighted by zone 1 and zone 2

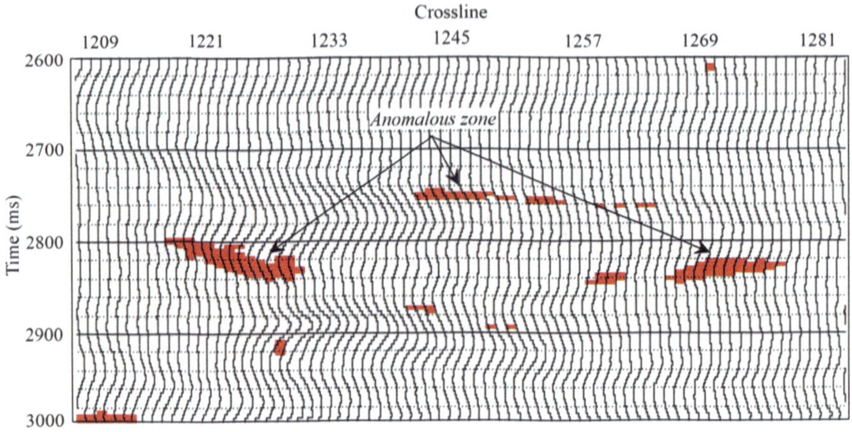

Fig. 4.22 Anomalous zones (red) are highlighted in the seismic section

seismic data and hence one can demonstrate the capability of these inversion methods to detect hydrocarbon zone (Figs. 4.20, 4.21 and 4.22).

References

Aki K, Richards PG (1980) Quantitative seismology. W.H. Freeman and Company, San Francisco, CA

Ankeny L, Braile L, Olsen K (1986) Upper crustal structure beneath the Jemez Mountains volcanic field, New Mexico, determined by three-dimensional simultaneous inversion of seismic refraction and earthquake data. J Geophys Res Solid Earth 91(B6):6188–6198

Buland A, Omre H (2003) Bayesian linearized AVO inversion. Geophysics 68(1):185–198

Buland A, Landre M, Andersen M, Dahl T (1996) AVO inversion of troll field data. Geophysics 61(6):1589–1602

Burianyk M, Pickfort S (2000) Amplitude-vs-offset and seismic rock property analysis: a primer. CSEG Recorder 25(9):6–16

Castagna JP, Batzle ML, Eastwood RL (1985) Relationships between compressional-wave and shear-wave velocities in clastic silicate rocks. Geophysics 50(4):571–581

Connolly P (1999) Elastic impedance. Lead Edge 18:438–452

Demirbag E, Coruh C, Costain JK, Castagna JP, Backus MM (1993) Inversion of P-wave AVO in offset-dependent reflectivity. Theory Practice AVO Anal Soc Expl Geophys 287–302

Fatti JL, Smith GC, Vail PJ, Strauss PJ, Levitt PR (1994) Detection of gas in sandstone reservoirs using avo analysis: a 3-d seismic case history using the geo-stack technique. Geophysics 59(9):1362–1376

Gardner GHF, Gardner LW, Gregory AR (1974) Formation velocity and density—the diagnostic basics for stratigraphic traps. Geophysics 39:770–780

Goodway B, Chen T, Downton J (1997) Improved AVO fluid detection and lithology discrimination using Lamé petrophysical parameters; "$\lambda\rho$" $\mu\rho$ and λ/μ fluid stack" From P and S inversions. In: SEG Annual Meeting, Society of Exploration Geophysicists, pp 183–186

Gray D, Andersen E (2000) The application of AVO and inversion to the estimation of rock properties. In: SEG Technical Program Expanded Abstracts, Society of Exploration Geophysicists, pp 549–552

Hampson D (1991) AVO inversion, theory and practice. The Leading Edge 10(6):39–42
Hampson DP, Russell BH, Bankhead B (2005) Simultaneous inversion of pre-stack seismic data. In: SEG Technical Program Expanded Abstracts, Society of Exploration Geophysicists, pp 1633–1637
Larson RG (1999) The structure and rheology of complex fluids, vol 150. Oxford University Press, New York
Lu S, McMechan GA (2004) Elastic impedance inversion of multichannel seismic data from unconsolidated sediments containing gas hydrate and free gas elastic inversion of gas hydrate. Geophysics 69(1):164–179
Ma XQ (2001) A constrained global inversion method using an overparameterized scheme: application to post-stack seismic data. Geophysics 66(2):613–626
Ma XQ (2002) Simultaneous inversion of prestack seismic data for rock properties using simulated annealing. Geophysics 67(6):1877–1885
Mallick S (2001) AVO and elastic impedance. Lead Edge 20(10):1094–1104
Mallick S (1995) Model-based inversion of amplitude-variations-with-offset data using a genetic algorithm. Geophysics 60(4):939–954
Maurya SP, Singh KH (2015) Reservoir characterization using model-based inversion and probabilistic neural network. Discovery 49(228):122–127
Maurya SP (2019) Estimating elastic impedance from seismic inversion method: a case study from Nova Scotia field, Canada. Curr Sci 116(4):1–8
Maurya SP, Singh NP (2018) Application of LP and ML sparse spike inversion with a probabilistic neural network to classify reservoir facies distribution—a case study from the Blackfoot Field, Canada. J Appl Geophys 159:511–521
Mora P (1987) Nonlinear two-dimensional elastic inversion of multi offset seismic data. Geophysics 52(9):1211–1228
Nolet G (1978) Simultaneous inversion of seismic data. Geophys J Int 55(3):679–691
Pan GS, Young CY, Castagna JP (1994) An integrated target-oriented pre-stack elastic waveform inversion: sensitivity, calibration, and application. Geophysics 59(9):1392–1404
Paul A, Cattaneo M, Thouvenot F, Spallarossa D, Béthoux N, Fréchet J (2001) A three-dimensional crustal velocity model of the southwestern Alps from local earthquake tomography. J Geophys Res Solid Earth 106(B9):19367–19389
Richards PG, Frasier CW (1976) Scattering of elastic waves from depth-dependent inhomogeneities. Geophysics 41(3):441–458
Russell B (1988) Introduction to seismic inversion methods. The SEG Course Notes, Series 2
Sen MK, Stoffa PL (1991) Nonlinear one-dimensional seismic waveform inversion using simulated annealing. Geophysics 56(10):1624–1638
Sherrill FG, Mallick S, WesternGeco L.L.C. (2008) 3D pre-stack full-waveform inversion. U.S. Patent 7,373,252
Shuey R (1985) A simplification of the zoeppritz equations. Geophysics 50(4):609–614
Shu-jin Y (2007) Progress of pre-stack inversion and application in exploration of the lithological reservoirs. Progr Geophys 3:032
Simmons JL, Backus MM (1996) Waveform-based avo inversion and avo prediction-error. Geophysics 61(6):1575–1588
Simmons JL, Backus MM (2003) An introduction multicomponent. Lead Edge 22(12):1227–1262
Tarantola A (1986) A strategy for nonlinear elastic inversion of seismic reflection data. Geophysics 51(10):1893–1903
Whitcombe DN (2002) Elastic impedance normalization. Geophysics 67(1):60–62

Chapter 5
Amplitude Variation with Offset (AVO) Inversion

Abstract In this chapter, the details of Amplitude variation with offset has been discussed. The method utilizes CMP gathers to produce AVO attribute which is used as good indicators of the presence of gas formation. The chapter is divided into two parts, in the first part it discusses about fluid replacement modeling and in the second section, it discusses about AVO inversion. Initially, the details of these methods are provided and thereafter, synthetic as well as real data examples are discussed.

5.1 Introduction

Variation in seismic reflection amplitude with a change in distance between the shot point and receiver which indicates differences in lithology and fluid content in rocks above and below the reflector can be used to analyze rock and fluid types in the reservoir. AVO analysis and inversion is a technique by which geophysicists attempt to determine thickness, porosity, density, velocity, lithology and fluid content of rocks. Successful AVO analysis requires special processing of seismic data and seismic modeling to determine rock properties with known fluid content. With that knowledge, it is possible to model other types of fluid content. A gas-filled sandstone might show increasing amplitude with offset, whereas coal might show decreasing amplitude with offset (Russell 1988).

AVO is the comparison of changes in seismic amplitude with the offset of the source traces. The AVO inversion is first suggested by Ostrander in 1984. Analysis of seismic amplitude versus offset (AVO) is a strong geophysical technique that is helpful in the detection of gas directly from seismic records. The standard assessment is as follows. First, the efficient elastic parameter of a hydrocarbon reservoir needs to be estimated using elastic reflective coefficient formulation; which models the reservoir as porous media and infers the porous parameters from the efficient elastic parameters. Strictly speaking, the reflection coefficient is an approximate approach to porous media. The primary goal of AVO assessment is to use standard surface seismic information to acquire subsurface rock characteristics. These rock characteristics can then help to determine lithology, a saturation of fluid, and porosity. In 1955, Koefoed also suggested analyzing the shape of the coefficient of reflection versus angle of

S. P. Maurya et al., *Seismic Inversion Methods: A Practical Approach*, Springer Geophysics, https://doi.org/10.1007/978-3-030-45662-7_5

incidence curve as a technique of lithology interpretation. The Knott power equations (or Zoeppritz equations) solution has shown that the energy reflected from an elastic boundary varies with the incident wave angle of incidence (Muskat and Meres 1940; Koefoed 1955, 1962). He also explained that the change in the coefficient of reflection with the angle of the incident depends on the difference in the Poisson ratio across an elastic border. Poisson's ratio is described as the transverse strain-to-longitudinal strain ratio (Sheriff 1973) and is associated with an elastic medium's P-wave and S-wave velocities. The poison's ratio can be expressed as follows.

$$\sigma = \frac{\frac{1}{2}\left(\frac{V_P}{V_S}\right)^2 - 1}{\left(\frac{V_P}{V_S}\right)^2 - 1} \tag{5.1}$$

The reflection coefficient with angle of incidence changes significantly changed which are used to discriminate between gas saturated sands and brine saturated sands hence AVO could be used to detect gas sands (Coulombe 1993). A limitation of AVO analysis is use of only P-energy which yields failure to get a unique solution, so AVO results are prone to misinterpretation (Pendrel and Dickson 2003). One common misinterpretation is the failure to distinguish a gas-filled reservoir from a reservoir having only partial gas saturation. To overcome the problem, AVO analysis using source-generated or mode-converted shear wave energy allows the differentiation of degrees of gas saturation. AVO analysis and inversion are more successful in young, poorly consolidated rocks than in older, well-cemented sediments (Simmons and Backus 1996).

The objective of this chapter is to understand rock properties at the possible reservoir zone, to do AVO (Amplitude vs. Offset) modeling, and to test the effectiveness of AVO modeling and analysis on the pre-stack seismic reflection data. We will use conventional velocity and amplitude analysis, and will also try to relate the hydrocarbon saturation to the amplitude or velocity variation. AVO inversion can be performed in two steps, first Fluid Replacement Modeling is carried out to see the changes in seismic signature due to fluid substitution in the reservoir zone and then AVO analysis is performed to estimate subsurface parameters.

5.2 Fluid Replacement Modeling

Fluid replacement modeling is a significant component of seismic attribute research as it provides the interpreter with a useful tool for modeling different fluid situations that could explain an observed variation in amplitude with offset anomaly (Smith et al. 2003). Modeling changes from one type of fluid to another involves first removing the effects of the starting fluid before modeling the fresh fluid. In practice, the rock is drained from its original pore fluid, calculating the porous frame's moduli (bulk and shear) and material density. Once correctly determined the porous frame characteristics, the rock is saturated with the fresh pore liquid, calculating the new efficient bulk moduli and density.

The most frequently used method is to apply the equations of Gassmann. The Gassmann equation relates the bulk modulus of the porous rock frame, mineral matrix and pore fluids (Batzle and Wang 1992). The use of Gassmann's equation is based on the following assumptions. The rock is homogeneous and isotropic, all pores are interconnected and interacting, pore pressure is balanced throughout the rock, the pore fluid does not communicate with the solid to soften or harden the frame, and the media is closed and no pore fluid leaves the volume of the rock (Smith et al. 2003; Wang 2001).

Changing the fluid type and its saturation value at the reservoir and creating synthetic logs relating to these changes is introduced as fluid replacement modeling. In most cases, only one well data is available in the study area which has encountered a specified horizon (oil, gas or water). In such a situation, modeling of AVO behavior for all conditions of the reservoir including gas, oil, and water is not possible. The solution is the modeling of reservoir rock and existing fluids, and then the prediction of synthetic logs with substituting fluid type using this modeling (Verm and Hilterman 1995). By having P-wave logs, shear wave, and density and using Zoeppritz's equations or one of its approximations, AVO behavior of reservoir can be estimated in different conditions of pore fluid and comparing that with seismic data through providing synthetic seismograph. Gassmann's (1951) equations are used to model reservoir rocks and fluid in almost all fluid modeling project. It is used to predict reservoir seismic properties changes (density, compressional and shear wave velocity) caused by fluid replacement based on texture property of reservoir rock. The Gassmann equation can be written as follows.

$$K_{sat} = K_{dry} + \frac{\left(1 - \frac{K_{dry}}{K_m}\right)^2}{\frac{\phi}{K_f} + \frac{1-\phi}{K_m} + \frac{K_{dry}}{K_m^2}} \tag{5.2}$$

In which K_{sat} is the bulk modulus of the rock saturated with pore fluid, K_{dry} is the bulk modulus of the dry frame (drained of any pore-filling fluid), K_m is the volumetric module resultant of forming minerals in reservoir rock and ϕ is reservoir rock porosity (Kumar 2006). The supporting equation can be written as follows.

$$K_{sat} = \rho_b\left(V_P^2 - \frac{4}{3}V_S^2\right) \tag{5.3}$$

$$\mu = \rho_b V_S^2 \tag{5.4}$$

The bulk modulus or incompressibility of anisotropic rock, K, is the ratio of hydrostatic stress to volumetric strain and is related to V_P, V_S, and ρ_b. Shear modulus or shear stiffness, μ is the ratio of shear stress to shear strain and is related to V_S, and ρ_b.

$$V_P = \sqrt{\frac{\left(K_{sat} + \frac{4}{3}\mu\right)}{\rho_b}} \tag{5.5}$$

$$V_S = \sqrt{\frac{\mu}{\rho_b}} \tag{5.6}$$

$$\rho_b = \rho_{ma}(1 - \phi) + \rho_{fl}\phi \tag{5.7}$$

In the above equations, the V_P represent P-wave velocity, V_S is S-wave velocity, ρ_b is the bulk density of the formation, ρ_{ma} is the density of the matrix and ρ_{fl} is the density of the fluid.

Smith et al. (2003) stated that while the K_{sat} of a rock may be sensitive to the composition of the pore fluid; the μ is insensitive and therefore does not vary in the course of fluid substitution.

$$\mu_{dry} = \mu_{wet} \tag{5.8}$$

The objective of fluid substitution is to model the seismic properties (seismic velocities) and density of a reservoir at a given reservoir condition (pressure, temperature, porosity, mineral type, and water salinity) and pore fluid saturation such as 100% water saturation or hydrocarbon with only oil or only gas saturation. The density of a saturated rock can be simply computed with the volume averaging equation (mass balance). Other parameters required to estimate seismic velocity after fluid substitution is the moduli which can be computed using the Gassmann's equations.

5.2.1 *Gassmann's Equations*

Gassmann's equations relate the bulk modulus of rock to its pore, frame, and fluid properties. Recall the Gassmann equation to estimate the bulk modulus of a saturated rock.

$$K_{sat} = K_{frame} + \frac{\left(1 - \frac{K_{frame}}{K_{matrix}}\right)^2}{\frac{\phi}{K_{fl}} + \frac{(1-\phi)}{K_{matrix}} - \frac{K_{frame}}{K_{matrix}^2}} \tag{5.9}$$

where, K_{frame}, K_{matrix}, and K_{fl} are the bulk moduli of the saturated rock and porous rock frame (drained of any pore-filling fluid), mineral matrix, and pore fluid, respectively, and ϕ is porosity (as fa reaction). In the Gassmann formulation, the shear modulus is independent of the pore fluid as discussed earlier and is constant during the fluid substitutions.

In order to estimate the saturated bulk modulus (Eq. 5.9) in a given condition of the reservoir and fluid type, it is necessary to estimate the bulk module of frame, matrix and pore fluid (Han and Batzle 2004). In the following section, we will address formulations for calculating bulk modulus and mineral matrix density, pore fluid, and rock frame.

5.2.2 *Fluid Properties*

Bulk module and pore fluid density (brine, oil, and gas) are estimated by an average of fluid type values. Let's first calculate each type of liquid (brine, oil, and gas) characteristics.

(i) *Estimation of Bulk modulus and density of brine*

The bulk brine module can be estimated from known seismic velocity and brine density. The related equation can be written as follows.

$$K_{brine} = \rho_{brine} V_{brine}^2 \times 10^{-6} \tag{5.10}$$

In Eq. 5.10, K_{brine} is bulk modulus in (GPa), ρ_{brine} is brine density in g/cm^3 and V_{brine} is the P-wave velocity in brine and can be taken as m/s (Han and Batzle 2004).

The brine density can be estimated using equation suggested by Batzle and Wang (1992). This can be written as follows.

$$\begin{aligned}\rho_{brine} &= \rho_w + 0.668S + 0.44S^2 \\ &\quad + 10^{-6}S[300P - 2400PS + T(80 + 3T - 3300S - 13P + 47PS)]\end{aligned} \tag{5.11}$$

In Eq. 5.11, ρ_w is water density presented in g/cm^3, S is the salinity of brine, P is pressure in situ given in MPa and T (°C) is temperatures (Wang 2001). The water density is largely dependent on temperature and pressure and can be written as follows.

$$\begin{aligned}\rho_w = 1 + 10^{-6}\big(&-80T - 3.3T^2 + 0.00175T^3 + 489P - 2TP + 0.016T^2P - 1.3 \\ &\times 10^{-5}T^3P - 0.333P^2 - 0.002TP^2\big)\end{aligned} \tag{5.12}$$

On the other hand, the P-wave velocity of brine can also be calculated using the formula suggested by Batzle and Wang (1992) and can be written as follows.

$$\begin{aligned}V_{brine} = V_w + S\big(&1170 - 9.6T + 0.055T^2 - 8.5 \times 10^{-5}T^3 + 2.6P - 0.0029TP \\ &- 0.0476P^2\big) + S^{1.5}\big(780 - 10P + 0.16P^2\big) - 1820S^2\end{aligned} \tag{5.13}$$

In Eq. 5.13, V_w shows the P-wave velocity in pure water. There is a relation between V_w, temperature (T) and pressure (P), and hence they can use to calculate P-wave velocity as follows.

Table 5.1 Coefficients for water velocity computation

$w_{11} = 1402.85$	$w_{13} = 3.437 \times 10^{-3}$
$w_{21} = 4.871$	$w_{23} = 1.739 \times 10^{-4}$
$w_{31} = -0.04783$	$w_{33} = -2.135 \times 10^{-6}$
$w_{41} = 1.487 \times 10^{-4}$	$w_{43} = -1.455 \times 10^{-8}$
$w_{51} = -2.197 \times 10^{-7}$	$w_{53} = 5.23 \times 10^{-11}$
$w_{12} = 1.524$	$w_{14} = -1.197 \times 10^{-5}$
$w_{22} = -0.0111$	$w_{24} = -1.628 \times 10^{-6}$
$w_{32} = 2.747 \times 10^{-4}$	$w_{34} = 1.237 \times 10^{-8}$
$w_{42} = -6.503 \times 10^{-7}$	$w_{44} = 1.327 \times 10^{-10}$
$w_{52} = 7.987 \times 10^{-10}$	$w_{54} = -4.614 \times 10^{-13}$

$$V_w = \sum_{i=1}^{5} \sum_{j=1}^{4} w_{ij} T^{i-1} P^{j-1} \tag{5.14}$$

In Eq. 5.14, w_{ij} present constant and given by Batzle and Wang (1992). These values are given in Table 5.1.

(ii) *Estimation of Bulk modulus and density of oil*

The oil contains some amount of gas dissolved in it and characterized by GOR (gas-to-oil ratio). The density of oil and the bulk modulus depend on the temperature, pressure, GOR and type of oil present. The oil density (ρ_{oil}) values can be estimated using the relationship provided by Batzle and Wang (1992) and Wang (2001) as follows.

$$\rho_{oil} = \frac{\rho_S + \left(0.00277P - 1.71 \times 10^{-7} P^3\right)(\rho_S - 1.15)^2 + 3.49 \times 10^{-4} P}{0.972 + 3.81 \times 10^{-4}(T + 17.78)^{1.175}} \tag{5.15}$$

where ρ_S is saturation density, P is pressure and T is temperature. The saturation density and depends on temperature and pressure and can be calculated as follows.

$$\rho_S = \frac{\rho_0 + 0.0012 R_G G}{B_0} \tag{5.16}$$

where ρ_0 is the reference density of oil measured at 15.6 °C and taken in (g/cm^3), R_G is the gas-oil-ratio (GOR) given in litre/litre, G is the specific gravity of oil in (API) and B_0 is called the formation volume factor and can be calculated as follows.

$$B_0 = 0.972 + 0.00038\left[2.49 R_G \sqrt{\frac{G}{\rho_0}} + T + 17.8\right]^{1.175} \tag{5.17}$$

Further, the P-wave velocity in oil (V_{oil}) can also be calculated by the relationship given by Batzle and Wang (1992) and Wang (2001) and the equation can be written as follows.

$$V_{oil} = 2096\sqrt{\frac{\rho_{ps}}{2.6 - \rho_{ps}}} - 3.7T + 4.64P + 0.0115\left[\sqrt{\frac{18.33}{\rho_{ps}}} - 16.97 - 1\right]TP \tag{5.18}$$

In Eq. 5.18, ρ_{ps} is pseudo density and can be calculated as follows.

$$\rho_{ps} = \frac{\rho_0}{(1 + 0.001R_G)B_0} \tag{5.19}$$

Using the relationship given above, one can estimate velocity and density of oil and hence, the bulk modulus of oil, K_{oil}(GPa) can be calculated as follows.

$$K_{oil} = \rho_{oil}V_{oil}^2 \times 10^{-6} \tag{5.20}$$

Fluid composed of brine and hydrocarbon (oil and/or gas) in the pore spaces and hence the bulk modulus and mixed pore liquid phase density can be estimated by reverse bulk modulus average and arithmetic density average of distinct liquid stages, respectively (Inyang 2009). Mathematically, the bulk modulus (K_{fl}) can be written as follows.

$$\frac{1}{K_{fl}} = \frac{WS}{K_{brine}} + \frac{HS}{K_{hyc}} \tag{5.21}$$

where WS is the water saturation (as fraction) and HS (=1 − WS) is the hydrocarbon saturation, K_{hyc} and ρ_{hyc} are the bulk modulus and density of hydrocarbon, respectively. Similarly, fluid density (ρ_{fl}) can be written as follows.

$$\rho_{fl} = WS\rho_{brine} + HS\rho_{hyc} \tag{5.22}$$

In the case of oil as hydrocarbon, the bulk modulus of hydrocarbon (K_{hyc}) is the same as the bulk modulus of oil and the density of hydrocarbon is the same as the density of oil (Kumar 2006). Similarly, in the case of gas, the bulk modulus of hydrocarbon is the same as the bulk modulus of gas and the density of hydrocarbon is the same as the density of the gas.

(iii) *Estimation of Bulk modulus and density of gas*

The bulk module and gas density in a reservoir depend on the pressure, temperature and gas type. Hydrocarbon gas can be a mixture of many gasses, characterized by specific gravity G, gas density-to-air density ratio at 15.6 °C, and atmospheric pressure. The approximate value of gas density can be calculated using the relationship provided again by Batzle and Wang (1992). The relationship can be written as follows.

$$\rho_{gas} \cong \frac{28.8GP}{ZR(T + 273.15)} \tag{5.23}$$

In Eq. 5.23, G represents the specific gravity of gas (API), R is the gas constant taken as 8.314 and Z is the compressibility factor which can be estimated using the relation between pseudo reduced temperature and pressure and is given as followings.

$$Z = \left[0.03 + 0.00527(3.5 - T_{pr})^3\right]P_{pr} + (0.64T_{pr} - 0.007T_{pr}^4 - 0.52) + 0.109(3.85 - T_{pr})^2 \exp\left\{-\left[0.45 + 8\left(0.56 - \frac{1}{T_{pr}}\right)^2\right]^{P_{pr}^{1.2}/T_{pr}}\right\} \tag{5.24}$$

In Eq. 5.24, T_{pr} represents pseudo reduced temperature, P_{pr} shows pseudo reduced pressure and can be calculated using the following formula.

$$T_{pr} = \frac{T + 273.15}{94.72 + 170.75G} \tag{5.25}$$

And

$$P_{pr} = \frac{P}{4.892 - 0.4048G} \tag{5.26}$$

Further, the bulk modulus of gas is suggested by Batzle and Wang (1992) and can be written as follows.

$$K_{gas} \cong \frac{P}{\left(1 - \frac{P_{pr}}{Z}\frac{\partial Z}{\partial P_{pr}}\right)_T} \frac{\left(0.85 + \frac{5.6}{P_{pr}+2} + \frac{27.1}{(P_{pr}+3.5)^2} - 8.7\exp\left[-0.65(P_{pr} + 1)\right]\right)}{1000} \tag{5.27}$$

In the above equation,

$$F = -1.2\frac{P_{pr}^{0.2}}{T_{pr}}\left[0.45 + 8\left(0.56 - \frac{1}{T_{pr}}\right)^2\right] \times \exp\left\{-\left[0.45 + 8\left(0.56 - \frac{1}{T_{pr}}\right)^2\right]^{P_{pr}^{1.2}/T_{pr}}\right\} \tag{5.28}$$

And

$$\left(\frac{\partial Z}{\partial P_{pr}}\right)_T = 0.03 + 0.00527(3.5 - T_{pr})^3 + 0.109(3.85 - T_{pr})^2 F \tag{5.29}$$

Using the above relationship, one can calculate the bulk modulus and density of the gas. Once these parameters are known, one can move to calculate matrix properties.

5.2.3 *Matrix Properties*

To calculate the mineral matrix's bulk modulus, one needs to understand the rock's mineral structure that can be discovered from the core sample laboratory examination. Lithology may be presumed to be a combination of quartz and clay minerals in the lack of laboratory information (Kumar 2006). The proportion of clay can be obtained from the curve of volume shale (V_{sh}), typically obtained from the information of the wireline log (Gamma-ray log). It can be noted that the shale formation contains about 70% of clay along with 30% quartz. Once the mineral abundances are determined one can calculate the bulk modulus of the matrix (K_{matrix}) using Voigt-Reuss-Hill (VRH) (Fig. 5.1) averaging formula (Hill 1952).

$$K_{matrix} = \frac{1}{2}\left(\left[V_{clay}K_{clay} + V_{qtz}K_{qtz}\right] + \left[\frac{V_{clay}}{K_{clay}} + \frac{V_{qtz}}{K_{qtz}}\right]\right) \quad (5.30)$$

where V_{sh}, K_{clay} and K_{qtz} is shale volume, the bulk modulus of clay and bulk modulus of quartz, respectively. These parameters can be calculated as follows.

$$V_{clay} = 70\% V_{qtz} \quad (5.31)$$

And

$$V_{qtz} = 1 - V_{clay} \quad (5.32)$$

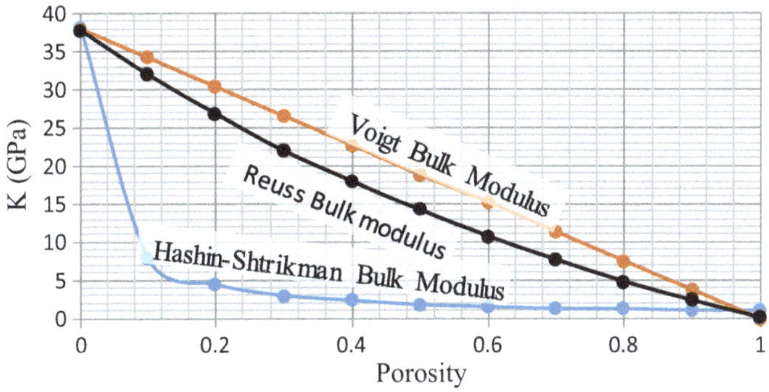

Fig. 5.1 Curves of Reuss (R), Voigt (V) and Hashin-Shtrickman (HS), showing bulk modulus variation with the porosity. Using Ko = 38 GPa and Go = 44 GPa, clean sandstone parameters (After Hashin and Shtrikman 1963)

Table 5.2 Rock properties and their standard value

Properties	Value
Bulk moduli of clay	20.9 GPa
Bulk moduli of quartz	36.6 GPa
Densities of clay	2.58 g/cm^3
Densities of quartz	2.65 g/cm^3

The density of the mineral matrix ρ_{matrix} can be estimated by arithmetic averaging of densities of individual minerals as

$$\rho_{matrix} = V_{clay}\rho_{clay} + V_{qtz}\rho_{qtz} \tag{5.33}$$

where ρ_{clay} and ρ_{qtz} are the density of the clay and quartz minerals. The standard value of these parameters can be found in Table 5.2 (Mavko and Mukerji 1998).

From the formula, one can notice that the K_{matrix} and ρ_{matrix} remain constant during the Gassmann fluid substitution.

5.2.4 *Frame Properties*

Laboratory measurement, empirical relationship or wireline log information can be used to derive frame bulk modulus. K_{frame} can be determined when interacting with wireline information by rewriting the Gassman equation (Zhu and McMechan 1990; Kumar 2006) as follows.

$$K_{frame} = \frac{K_{sat}\left(\frac{\phi K_{matrix}}{K_{fl}} + 1 - \phi\right) - K_{matrix}}{\frac{\phi K_{matrix}}{K_{fl}} + \frac{K_{sat}}{K_{matrix}} - 1 - \phi} \tag{5.34}$$

All the parameters in Eq. (5.34) are known from the previous formulations and hence one can estimates K_{frame} which remain unchanged during fluid substitution.

5.2.5 *Application of FRM*

Seven layers have been classified from the well 01-17 of the Blackfoot field, Canada and the sixth layer (Layer 5) is interpreted as sandstone formation (Table 5.3). The fluid substation is performed for the sandstone layer. Table 5.4 shows list of constantly used as input in fluid replacement modeling.

From the analysis, it is noticed that the P-wave velocity drops steeply between 0 and 20% saturation and then increases slowly (Fig. 5.2a). The maximum decrease in P-wave velocity is found to be 2.7%. On the other hand, the S-wave velocity

Table 5.3 The layering of well 01-17 from Blackfoot field, Canada and corresponding velocity, density and porosity values

Layer	P-velocity (m/s)	S-velocity (m/s)	Density (kg/m^3)	Phi
Ocean (0–162 m)	1480	0	1000	NA
Layer 1 (162–510)	2978	1396	2060	0.36
Layer 2 (510–812)	3248	1628	2150	0.30
Layer 3 (812–1332)	3515	1857	2310	0.20
Layer 4 (1332–1545)	3963	2244	2520	0.08
Layer 5 (1545–1584)	3932	2217	2460	0.12
Layer 6 (1584–1600)	4690	2871	2520	0.08

Table 5.4 Summary of constant later used in reservoir and calculating of acoustic velocities

Constants	Value
Porosity (φ)	0.11
Temperature	55.5 °C
The density of CO_2 (ρ_{CO_2})	800 kg/m^3
The density of water (ρ_w)	1020 kg/m^3
Density of matrix (ρ_m)	2650 kg/m^3
The bulk modulus of CO_2 (K_{CO_2})	0.136 GPa
The bulk modulus of water (K_w)	2.39 GPa
The bulk modulus of matrix/solid (K_S)	37 GPa
The bulk modulus of dry rock (K_d)	2.56 GPa
Shear modulus of dry rock (μ_d)	0.8569 GPa

increases with CO_2 saturation and shows direct proportionality (Fig. 5.2b). The average increase of the S-wave velocity with respect to 100% water saturation is 2.44%. The other parameter analysed here are density which decreases with increasing CO_2 saturation (Fig. 5.2c). The average decrease of density with CO_2 saturation is found to be 4.73% as compared with 100% water saturation. Further, V_P/V_S the ratio is analyses and found that the V_P/V_S ratio decreases with increasing CO_2 saturation (Fig. 5.2d). This can be expected as the S-wave velocity increase and P-wave velocity decrease. The maximum drop-in V_P/V_S ratio found to be 7.75% in comparison with 100% water saturation. Figure 5.2d shows variation of V_P/V_S ratio and depicts a sharp drop between 0 and 20% CO_2 saturation and thereafter it becomes nearly constant (Table 5.5).

Further, the effect of this fluid substitution can also be monitored from the seismic amplitude variation. As we have already seen that the velocity and density changes due to fluid substitution and hence these petrophysical changes are expected to be reflected in the seismic response. In this regard, the synthetic seismograms are generated using forward modeling procedure from the velocity and density of the well log 08-08. The forward modeling is explained briefly in the following paragraphs.

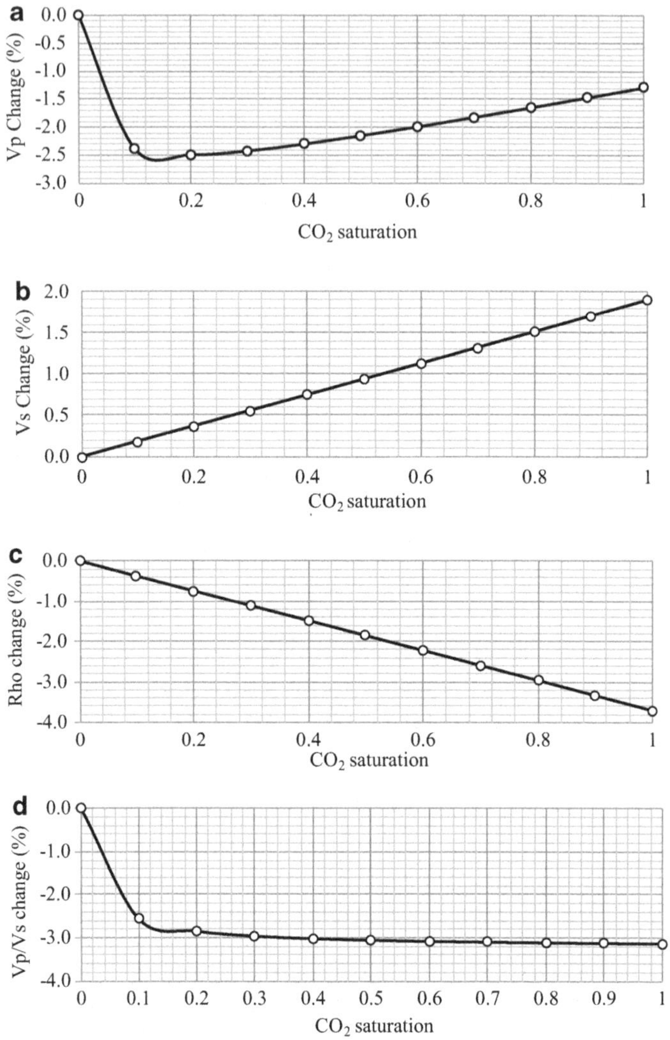

Fig. 5.2 **a** P-wave velocity change, **b** S-wave velocity change, **c** density change, **d** Vp/Vs changes with CO_2 saturation

5.3 AVO Modeling

After fluid submission, pre-stack synthetic seismograms are generated. The angle-dependent reflection coefficient ($R_{pp}(\theta)$) are calculated following Shuey's (1985) three-term approximation.

$$R_{pp}(\theta) \approx R(0) + G\sin^2\theta + F\left(\tan^2\theta - \sin^2\theta\right) \tag{5.35}$$

Table 5.5 Gassmann fluid substitution results and input values (porosity, P-wave velocity, S-wave velocity, bulk modulus, Vp/Vs and shear modulus)

Water Sat	CO_2 Sat	V_P	V_S	ρ	V_P change	V_S change	ρ change	V_P/V_S change
1	0.0	3969.9	2179.2	2.44	0.0	0.00	0.00	0.00
0.9	0.1	3875.5	2183.3	2.43	−2.4	0.19	−0.37	−2.56
0.8	0.2	3872.6	2187.5	2.42	−2.5	0.37	−0.74	−2.85
0.7	0.3	3873.8	2191.4	2.41	−2.4	0.56	−1.11	−2.96
0.6	0.4	3878.7	2195.5	2.40	−2.3	0.75	−1.48	−3.02
0.5	0.5	3884.5	2199.7	2.39	−2.1	0.94	−1.85	−3.06
0.4	0.6	3890.8	2203.8	2.38	−2.0	1.13	−2.22	−3.08
0.3	0.7	3897.5	2208.0	2.37	−1.8	1.32	−2.59	−3.10
0.2	0.8	3904.4	2212.2	2.36	−1.6	1.51	−2.96	−3.12
0.1	0.9	3911.4	2216.5	2.35	−1.5	1.71	−3.33	−3.13
0.0	1.0	3918.6	2220.7	2.35	−1.3	1.90	−3.70	−3.14

Depth of 1545 m in the well site. Reservoir thickness 39 m

With normal incidence, $R(0)$, is defined as follows.

$$R(0) = \frac{1}{2}\left(\frac{\Delta V_p}{V_p} + \frac{\Delta \rho}{\rho}\right) \tag{5.36}$$

And the variation of the reflectivity versus the angle, the AVO gradient, G can be calculated as follows.

$$G = R(0) - \frac{\Delta V_\rho}{V_\rho}\left(\frac{1}{2} + \frac{2\Delta V_s^2}{V_s^2}\right) - \frac{4\Delta V_s^2}{V_p^2}\frac{\Delta V_s}{V_s} \tag{5.37}$$

And F, the reflectivity at the far angles (reflection angles higher than 30°) can be expressed as follows.

$$F = \frac{1}{2}\frac{\Delta V_p}{V_p} \tag{5.38}$$

Each elastic property is defined on each side of the interface where the reflection is happening as follows.

$$\Delta V_p = V_{p2} - V_{p1}$$
$$V_p = \frac{V_{p2} + V_{p1}}{2}$$
$$\Delta V_s = V_{s2} - V_{s1}$$

$$V_s = \frac{V_{s2} + V_{s1}}{2}$$
$$\Delta V_\rho = V_{\rho 2} - V_{\rho 1}$$
$$V_\rho = \frac{V_{\rho 2} + V_{\rho 1}}{2}$$

Index 1 refers to the vertical location at which the calculation of the reflection coefficient is performed, the mean above the reflection interface; while index 2 refers to the sample immediately below, the mean below the reflection interface (Shuey 1985). Figure 5.3 shows the process of the generation of the synthetic trace.

By using the linear approximation of Shuey within a geostatistical inversion approach, we can obtain the AVO normal-incidence, $\boldsymbol{R}(0)$ and gradient, $\boldsymbol{G}$, cubes immediately and as part of the inverse solution (Rutherford and Williams 1989; Avseth and Bachrach 2005; Castagna and Backus 1993; Maurya and Singh 2019). The propagation of uncertainty directly from the inverse problem to the inverted volumes of AVO allows for better risk assessment of interest anomalies in amplitude, which may be related to actual accumulations of hydrocarbons. It is crucial to note, however, that any alternative approximation can be used to calculate the coefficients of angle-dependent reflection.

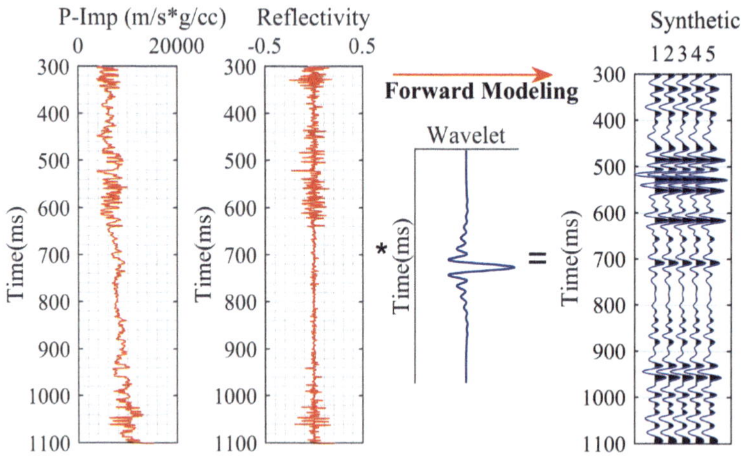

Fig. 5.3 Forward modeling procedure where the first track shows impedance, the second track shows reflectivity, third track shows wavelet and fourth track shows synthetic seismograms

5.3.1 Practical Aspect of FRM

The fluid replacement modeling has been performed for well log data (Penobscot L-30) from the Penobscot field, Scotian Shelf, Canada. The properties of fluid replacement modeling are as follows (Table 5.6).

With all the above parameters, AVO modeling is performed for Penobscot data, Canada and first, variation in rock properties like velocity, density, poison's ratio is monitored and then seismic amplitude variation. Three saturations have been modeled, the first saturation is taken as pure brine, the second saturation is pure gas and the third saturation has been taken as pure oil. Figure 5.4 shows variation of rock properties due to fluid substitution. It is noticed that the average change in P-wave velocity in the target zone is as follows. The change in velocity has been monitored by replacing reservoir fluid by pure brine, pure gas, and pure oil case one by one respectively. The 20% decrease has been noticed for pure brine case, a 23% decrease

Table 5.6 Input fluid and rock parameters of the formation used for AVO Modeling

Zone of Interest		8654.9 to 8756.6 ft	
Porosity threshold		2%	
Input fluid parameters			
Fluid parameters:	Brine	Fluid Parameters:	CO_2
Density:	1.09 g/cc	Density:	0.13 g/cc
Bulk modulus	2.38 GPa	Bulk modulus	0.041 GPa
Saturation	50%	Saturation	50%
Input matrix properties			
Quartz	100%	Gas bulk modulus	0.021 GPa
Density	2.65 g/cc	Gas density	0.1 g/cc
Bulk modulus	36.6 GPa	Brine bulk modulus	2.38 GPa
Shear modulus	45 GPa	Brine Density	1.09 g/cc
Output fluid parameters			
Fluid parameters	Brine	Fluid parameters	Oil
Density	1.09 g/cc	Density	0.75 g/cc
Bulk modulus	2.38 GPa	Bulk modulus	1 GPa
Saturation	50%	Saturation	0%
Fluid parameters	Gas		
Density	0.1 g/cc	Saturation	50%
Bulk modulus	0.021 GPa		

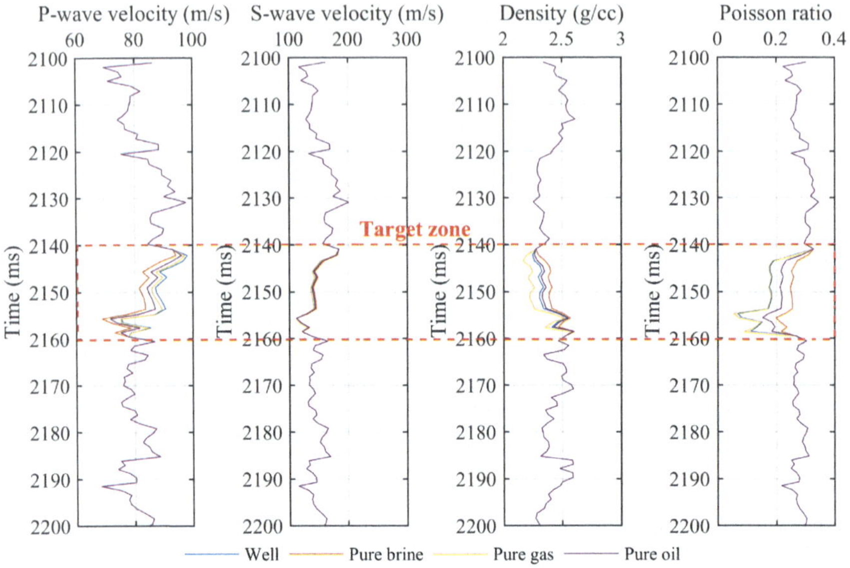

Fig. 5.4 Variation of rock properties due to fluid substitution

has been noticed for the pure gas case and a 27% decrease has been noticed for the pure oil case respectively.

5.3.2 *Direct Hydrocarbon Indicator*

A Direct Hydrocarbon Indicator is a seismic information characteristic that provides evidence of hydrocarbon existence and is particularly helpful in decreasing the risk of well drilling to a dry exploration (Mocnik 2011). It is critical that time is spent on choosing what the behaviors of the rock characteristics are and their manifestations before interpretation of the seismic section. As the acoustic impedance between a reservoir and the surrounding rock changes, one can notice a distinct type of seismic response. They can be like a bright spot, dim spot, flat spot, polarity reversal and Gas chimneys (Swan et al. 1993; Dey-Sarkar et al. 1995; Luping 2008; Tai et al. 2009).

The first use of amplitude information as hydrocarbon indicators was in the early 1970s when it was found that bright spot amplitude anomalies could be associated with hydrocarbon traps.

i. **Bright spot**

Hammond (1974) proposed the method of quantitative analysis of seismic data and revolutionized the search for gas and oil worldwide. This finding improved interest in rock physical characteristics and understanding of modifications in amplitude with distinct rock and pore fluid kinds (Gardner et al. 1974). It has been researched in comparatively smooth sand that the existence of gas dramatically improves the rock's compressibility, the velocity falls and the amplitude reduces producing a negative polarity, which explains the word bright spot. A seismic amplitude anomaly of high amplitude that could indicate the presence of hydrocarbons is called a bright spot. Bright spots are the result of major changes in acoustic impedance and effect of tuning, such as when gas sand underlies shale, but can also be caused by occurrences other than the presence of hydrocarbons, such as a lithological change (White 1977; Mocnik 2011). The term is frequently used as a hydrocarbon indicator. These amplitude peaks can be caused by a gas resulting in increase in the reflection coefficient in the pore space. It is used to define the rise in amplitude rather than the presence of hydrocarbons, and in unconsolidated clastic rocks, it is generally higher. As shown in Fig. 2.1, the reduced sand impedance generates an amplitude increase over the crest of the hydrocarbon structure.

ii. **Flat spot**

The flat spot depicts a seismic reaction of hydrocarbon contact where it appears to be flat. Such contact can be between gas and oil, between oil and water, or between gas and water. To depict a flat spot, the hydrocarbon reservoir is much thicker than the vertical resolution. Flat spots are often hard to locate; channel edge or base, low angle flaws, or artifacts processing can often be misunderstood as flat spots (Luping 2008; Mocnik 2011). Low saturated gas in a reservoir can also cause flat spots. Finding the so-called flat spot at the hydrocarbon-water touch is very crucial. This is a difficult reflector (increase in impedance) and should be on the same TWT as the shift in amplitude. If both oil and gas are present, two separate flat spots should be apparent, one at the contact between gas and oil and one closer to the contact between oil and water (Backus and Chen 1975; Swan et al. 1993).

iii. **Dim spot**

Highly consolidated sands with much higher acoustic impedance than the overlying shale are causing dim spots. As seen in Fig. 5.5, there is a strong peak at the top of the sandstone. Adding hydrocarbons to the pore space can cause the sandstone's velocity and density to reduce; however, the polarity of the reflection coefficient will not reduce sufficiently (Chen and Sidney 1997). The hydrocarbon decreases the impedance of acoustics and the coefficient of reflection, thus producing a dim spot. This situation can often be hard to interpret with confidence, especially if minor fault affects the structure. We expect to see a "dim place" or a decline in amplitude at the top reservoir for difficult brine sand relative to the shale and difficult hydrocarbon sand relative to the rock (Fig. 5.5). A dim spot is not simply to acknowledge and should be in line with the flat spot's structure and TWT (Mocnik 2011; Brown 2012).

Therefore, the elements to consider during the analysis of seismic information are several and mostly rely on the acoustic effects of a gas accumulation, the overlying material, the porosity, the depth, the water saturation and the configuration of the reservoir. Thus, the only observation of an amplitude anomaly may not be adequate to attribute a hydrocarbon origin; it is very essential to consider all the impacts on

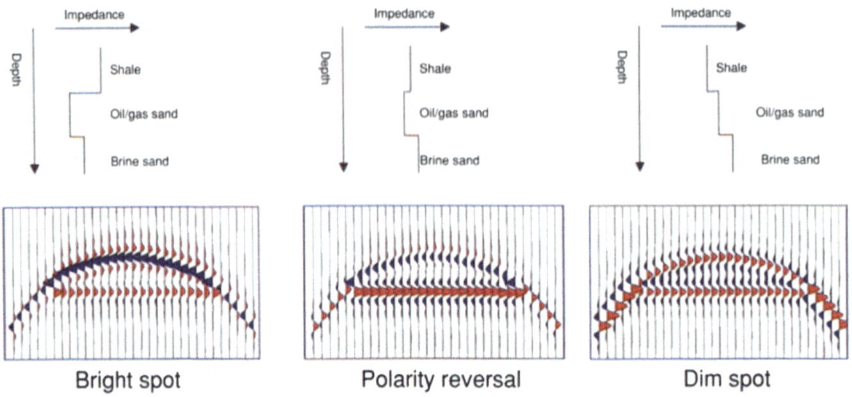

Fig. 5.5 Schematic model of a bright spot, polarity reversal and dim spot for different oil/gas brine sand responses (From Bacon et al. 2007)

the seismic signal generated by gas/oil presence, including, in specific, its elements, which is not only amplitude but also frequency and phase (Li and Mueller 1997).

iv. **Polarity reversal**

If the reservoir material in its water-saturated state has an acoustic impedance greater than the surrounding material, and if the local water-to-gas replacement depresses the acoustic impedance of the gas saturated zone to less than that of the surrounding material, then the reflection from the top of the reservoir shows a reversal of polarity over gas. This is not needed proof for gas as these circumstances may or may not occur; it may also be hard to identify polarity reversal owing to interferences and tiny fault impacts (Keys 1989; Chen and Sidney 1997; Luping 2008).

Phase change, also recognized as polarity reversal, happens when the surrounding reservoir has a reduced reservoir rock velocity. This can happen when a partly consolidated sand becomes moist. This leads to mildly greater acoustic impedance than the overlying shale. The bottom of the sand correlates with a small favorable coefficient of reflection. The velocity and density of the sandstone decrease as hydrocarbons are introduced to the pore space; which also results in a reduction in the acoustic impedance to the stage where it is mildly less than the overlying shale (Tai et al. 2009; Mocnik 2011). The reflection at the base then switches stage, from a maximum to a soft trough, which can be seen in Fig. 5.5.

5.3.3 Synthetic Modeling of AVO from Logs

Chiburis et al. (1993) indicated that the use of AVO for identifying fluid is achieved by using AVO comparing the actual information with a synthetic seismogram. Hampson-Russell's AVO is used in this study to generate AVO synthesized seismograms for fluid-saturated soils (petroleum, brine, and gas) with input density and recorded velocity (P-wave and S-wave) from well L-30 of Penobscot field, Canada. With the use of Zoeppritz and elastic flow equations, synthetics are produced for interval 2100–2300 ms and analytical outcomes are contrasted. While both equations calculate the amplitudes of seismic waves, the equations of Zoeppritz recognize only plane-wave amplitudes of reflected P-waves and disregard intermittent multiples and mode-converted waves (Russell 1999). On the other side, the elastic wave algorithm designs multiple waves as well as converted mode waves. Synthetics are developed using both algorithms taking into consideration failures in transmission and distributing geometry. These would lead to a fake Class III AVO. The seismic event of concern is the elevated amplitude corresponding to the boundary between reservoir sand and enclosing shales. Off the three synthetic seismograms (oil, brine, and gas) all show large amplitudes of normal incidents, the gas synthetics show the largest amplitude, brine shows the smallest, and oil falls between brine and gas.

These hydrocarbon indicators can be observed by the modeling of gas, oil, and brine in the reservoir. We have taken three cases, first, the reservoir is saturated with pure brine and corresponding changes in seismic sections are noticed. Thereafter, in the second and third cases, pure oil and pure gas substitution have been performed and corresponding seismic amplitude changes are noticed. Further, a comparison of seismic amplitude with offset in all three cases is also performed. These changes are provided in Figs. 5.6, 5.7 and 5.8.

From fluid replacement, it can be seen that the impact of gas saturation in the reservoir is greater than the same quantity of oil. The gas-filled reservoir rocks have the lowest impedance and the largest amplitudes of reflection, while brine rocks have the highest impedance and the lowest amplitudes of reflection. On the other side, oil sands have characteristics that lie between the characteristics of gas and brine.

5.4 Amplitude Variation with Offset (AVO) Analysis

Amplitude versus offset (AVO) or amplitude variation with offset is a variation in seismic reflection amplitude with a change in distance between the shot point and receiver that indicates differences in lithology and fluid content in rocks above and below the reflector. It is also referred to as amplitude variation with angle (AVA). AVO studies are generally performed in CMP gathers and indicate increases of offset with the angle. An AVO anomaly is most commonly expressed as increasing (rising) AVO in a sedimentary section, often where the hydrocarbon reservoir has lower acoustic impedance than the surrounding shale (Fatti et al. 1994; Dufour et al. 1998, 2002). Typically, the amplitude decreases with an offset due to geometrical spreading, attenuation, and other factors.

Typically, variation in seismic amplitude with variation in range between origin and receiver is connected with changes in lithology and fluid content in rocks above and below the reflector. But unique seismic data acquisition, processing, and interpretation methods are needed to achieve optimum outcomes from AVO analysis.

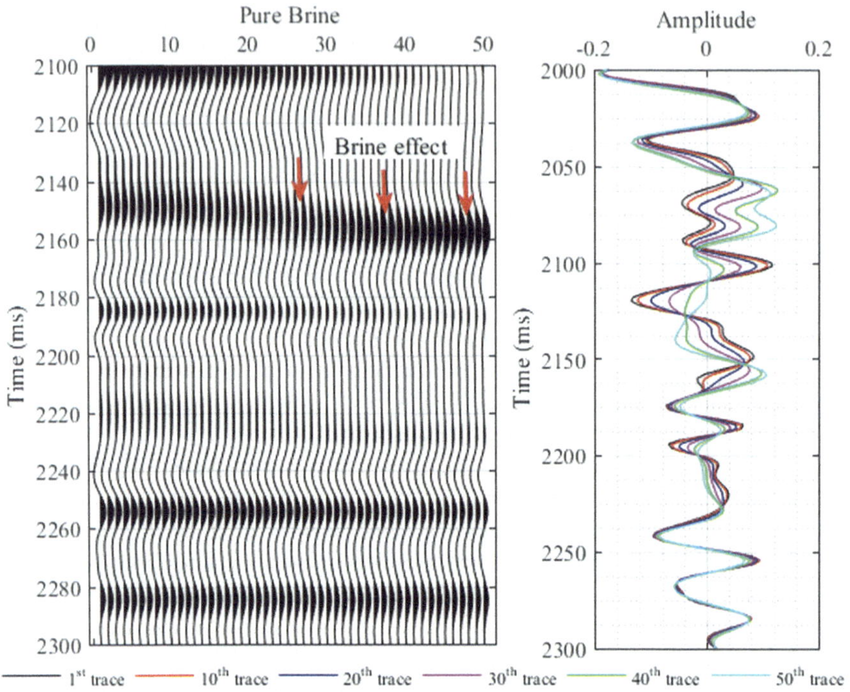

Fig. 5.6 Variation of seismic amplitude with offset for pure brine (left) and comparison of seismic amplitude (right)

The subsurface of the earth is quite complicated, so that distinct rocks have distinct reactions to AVO, yet these rocks are filled with the same fluid or have the same porosity.

5.4.1 Classification of AVO

Rutherford and Williams (1989) developed the classification scheme for AVO anomalies and altered it by Ross and Kinman (1995) and Castagna and Swan (1997).

Class 1: High-impedance contrast with decreasing AVO. The layer has a higher impedance than the surrounding shales.
Class 2: Near-zero impedance contrast between the sand and surrounding shales.
Class 2p: Same as class 2, but with polarity change;
Class 3: Low impedance with increasing AQVO compared to surrounding shales.
Class 4: Low impedance sand with decreasing AVO.

Graphically, it can be represented as in Fig. 5.9. Figure 5.9 shows plot of the

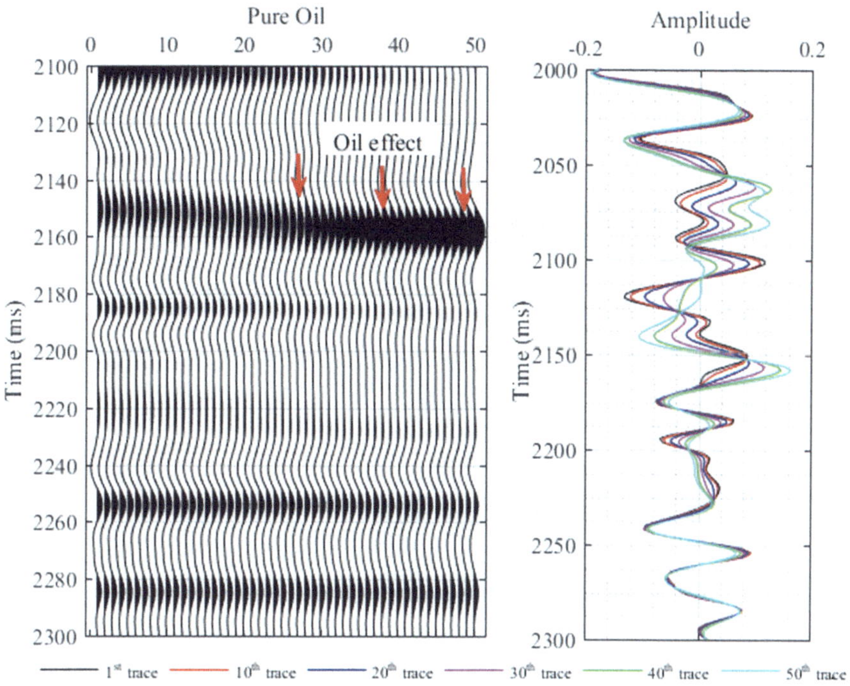

Fig. 5.7 Variation of seismic amplitude with offset for pure oil case (left) and comparison of seismic amplitude (right)

reflection coefficient with the incident angle. The four different kinds of AVO class is defined. These classifications are used to detect gas-bearing formation from no gas formations.

5.4.2 Theoretical Aspect of AVO

Basic AVO theory is well understood because it is widely used as a tool in hydrocarbon detection (Smith and Gidlow 1987). A few of the most important ideas will be highlighted to keep in mind when doing AVA analysis. The AVO can be understood by study the partition of energy from an interface. Figure 5.10 shows the theoretical energy partition at a boundary. This figure illustrates an important point that accounts for AVA phenomena: the conversion of P-wave energy to S-wave energy. Though the majority of seismic data is recorded simply as a single component pressure wave, the fact that the Earth is elastic causes amplitudes of P-wave arrivals to be a function of S-wave reflection coefficient (R_s) (Castagna and Smith 1994). In practice, R_s is tricky to obtain and the P-wave reflection coefficient (R_p) is what we have in the vast majority of cases (Smith and Sutherland 1996).

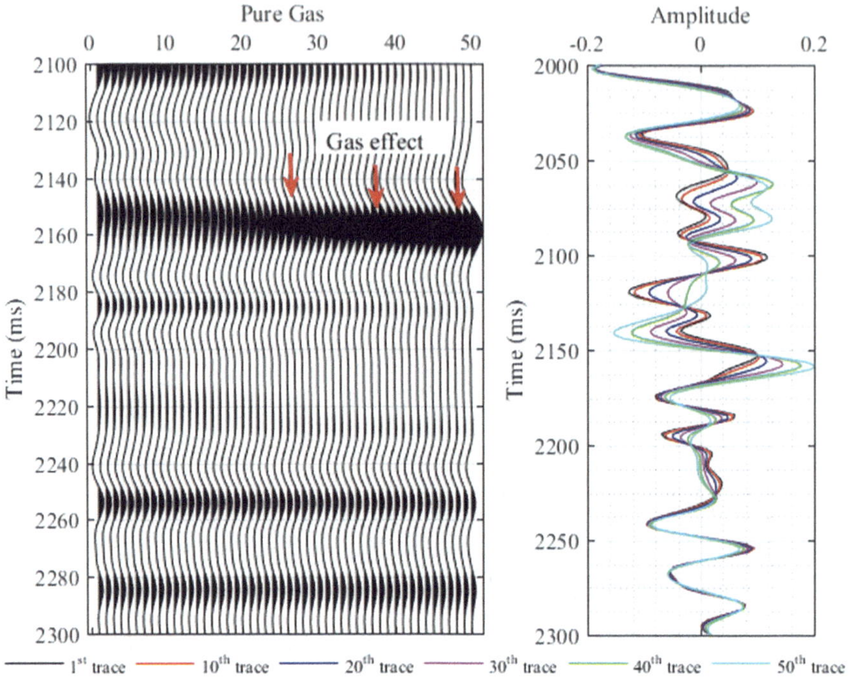

Fig. 5.8 Variation of seismic amplitude with offset for pure gas case (left) and comparison of seismic amplitude (right)

Using Snell's law, Knott in 1899, and Zoeppritz in 1919, developed general expressions for the reflection of compressional and shear waves at a boundary as a function of the densities and velocities of the layers in contact (Knott 1899; Zoeppritz 1919). Although Zoeppritz was not the first to publish a solution, his name is associated with the cumbersome set of formulas describing the reflection and refraction of seismic waves at an interface (Aki and Richards 1980).

i. **Zoeppritz equation**

The relationship between the incident and reflection/transmission amplitudes for plane waves at an elastic interface is described by the Zoeppritz equations. These equations give us the exact amplitudes as functions of the incidence angle. Figure 5.10 explains partitioning of energy of a boundary.

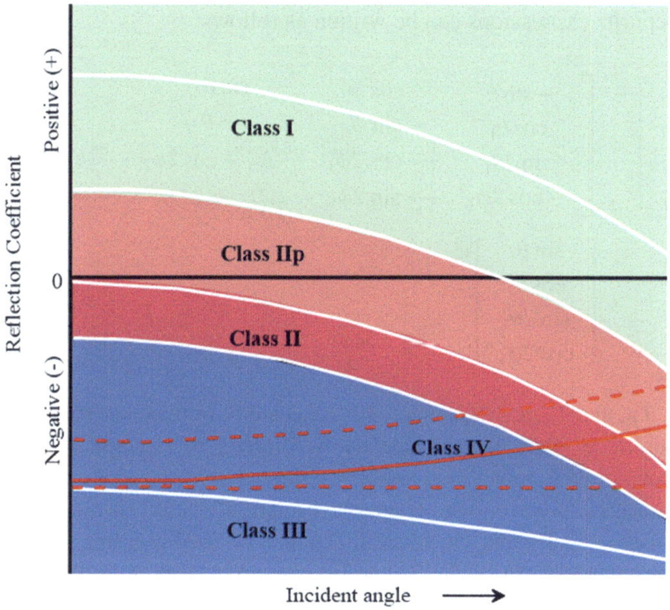

Fig. 5.9 AVO classifications for the top of gas sands

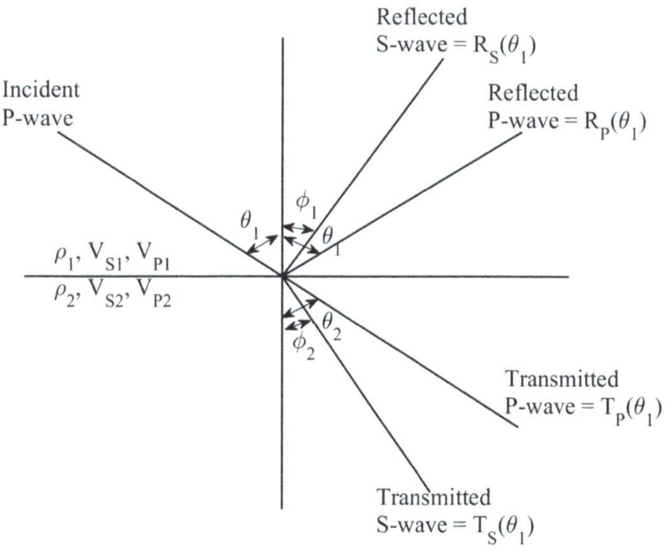

Fig. 5.10 Concept of the partitioning of energy from a boundary

The Zoeppritz expressions can be written as follows.

$$\begin{bmatrix} R_p(\theta_1) \\ R_S(\theta_1) \\ T_P(\theta_1) \\ T_S(\theta_1) \end{bmatrix} = \begin{bmatrix} -\sin\theta_1 & -\cos\phi_2 & \sin\theta_2 & \cos\phi_2 \\ \cos\theta_1 & -\sin\phi_1 & \cos\theta_2 & -\sin\phi_2 \\ \sin 2\theta_1 & \frac{V_{P1}}{V_{S1}}\cos 2\phi_1 & \frac{\rho_2 V_{S2}^2 V_{P2}}{\rho_1 V_{S1}^2 V_{P1}}\cos 2\phi_2 & \frac{\rho_2 V_{S2} V_{P1}}{\rho_1 V_{S1}}\cos 2\phi_2 \\ -\cos 2\phi_1 & \frac{V_{S1}}{V_{P1}}\sin 2\phi_1 & \frac{\rho_2 V_{P2}}{\rho_1 V_{P1}}\cos 2\phi_2 & -\frac{\rho_2 V_{S2}}{\rho_1 V_{P1}}\sin 2\phi_2 \end{bmatrix} \times \begin{bmatrix} \sin\theta_1 \\ \cos\theta_1 \\ \sin 2\theta_1 \\ \cos 2\phi_1 \end{bmatrix} \tag{5.39}$$

R_p, R_S, T_P, and T_S, are the reflected P, reflected S, transmitted P, and transmitted S-wave amplitude coefficients, respectively. Further, θ_1 is angle of incidence, θ_2 angle of transmitted P-wave, ϕ_1 is angle of reflected S-wave and ϕ_2 represents angle of the transmitted S-wave. The Zoeppritz Equation is used to estimate reflection and transmission coefficient, however, it can also be used as a function of angle by inverting the matrix form of the Zoeppritz equations (Larsen et al. 1999; Xu and Bancroft 1997).

The AVA/AVO analysis typically uses the small-contrast approximations to Zoeppritz equations. These approximations are discussed one by one in the following section.

ii. **Aki-Richards approximation**

The Zoeppritz equations discussed above were further linearized with respect to small variations of elastic parameters across the boundary, yielding an approximation of the full Zoeppritz equations (Aki and Richards 2002). This equation is termed as Aki-Richards approximation of Zoeppritz equations and can be expressed as follows.

$$R(\theta) = a\frac{\Delta P}{V_P} + b\frac{\Delta V_S}{V_S} + c\frac{\Delta\rho}{\rho}, \tag{5.40}$$

where $a = \frac{1}{2cos^2\theta}$, $b = 0.5 - \left(\frac{V_S}{V_P}\right)^2 \sin^2\theta$, and $c = 4\left(\frac{V_S}{V_P}\right)^2 \sin^2\theta$. Further, V_P represents p-wave velocity and V_S presents S-wave velocity, ρ is density and R is a reflection coefficient.

iii. **Wiggins approximation**

Wiggins et al. (1983) separated the three terms of the Zoeppritz equations related to perturbations in the elastic parameters of interest (Russell 1988). This can be written as follows.

$$R(\theta) = A + B\sin^2\theta + C\tan^2\theta\sin^2\theta \tag{5.41}$$

In Eq. 5.41, $A = \frac{1}{2}\left[\frac{\Delta V_P}{V_P} + \frac{\Delta \rho}{\rho}\right]$, $B = \frac{1}{2}\frac{\Delta V_P}{V_P} - 4\left[\frac{V_S}{V_P}\right]^2 \frac{\Delta V_S}{V_S} - 2\left[\frac{V_S}{V_P}\right]^2 \frac{\Delta \rho}{\rho}$, and $C = \frac{1}{2}\frac{\Delta V_P}{V_S}$.

These equations (Eqs. 5.40 and 5.41) predict an approximately linear relationship between the amplitude and $\sin^2\theta$ (Aki and Richards 2002). In Eq. 5.6, the intercept (A) is the zero-offset reflection coefficient, which is a function of the P-wave velocity and density. The AVO gradient (B) depends on the P- and S-wave velocities and density. The gradient has the largest effect on the amplitude variation with offset. The curvature factor (C) has only a very small effect on the amplitudes at incidence angles below 30°. By using the two terms of the Aki and Richards equation, one can extract them at different reflection times from the seismic amplitudes in CMP gathers. As a result, the intercept and gradient seismic attributes, $A(t)$ and $B(t)$ are produced.

In 3D seismic datasets, attributes A and B discussed above can be used to produce attribute volumes. However, such volumes are rarely used singly because they still do not provide unambiguous indicators of reservoir properties. Different combinations of these parameters are used to produce secondary attributes. Some of these combinations are discussed in the following sections.

1. AVO product ($A * B$)-This is a useful measure of classical bright spots where there are simultaneously high amplitudes (A) and enhanced gradients (B) occur (Castagna and Smith 1994). Class III gas sandstones, for instance, have small impedance, so both A and B are negative at the top and positive at the bottom. Consequently, for the top and bottom of such a reservoir, the product (A*B) indicates large positive values.
2. Scaled Poisson's Ratio Change ($A + B$)-This characteristic is based on the assumption that the Poisson ratio (σ) is approximately 1/3, and therefore A + B would be equivalent to (9/4) $\Delta\sigma$. Therefore, this characteristic is equal to the change in the proportion of the Poisson and therefore represents the changes in the proportion of V_P/V_S. In gas sand, it reduces at the top and rises at the bottom of the reservoir due to powerful variations in P-wave velocities coupled with only slight modifications in S-wave velocities.
3. Shear Reflectivity ($A - B$)-If we approximate, as it is commonly done, that $V_P/V_S = 2$, (corresponding to the Poisson's ratio of 1/3), then this attribute ($A - B$) is proportional to the shear-wave reflectivity ($2R_S$).

a. **AVO gradient analysis**

The method of gradient analysis enables parameters of the AVO function to be tested on a chosen seismic data collection. Therefore, this feature can be used to see what attributes to use in the volume and map functions of the AVO attribute. It also enables one to determine an AVO anomaly category (Ramos and Davis 1997).

The chart produces the Intercept when one models the amplitude of the wave for a reflector against the trace offset. This is where the zero-offset line meets the trend of amplitude measurements. The gradient, which is the curve slope produced by the plot points, is also produced (Li et al. 2007). It is then possible to use the sums or differences of these gradients and intercept values to map AVO anomalies. The offset is traditionally described as the sine squared of the angle of the event (Figs. 5.11 and 5.12).

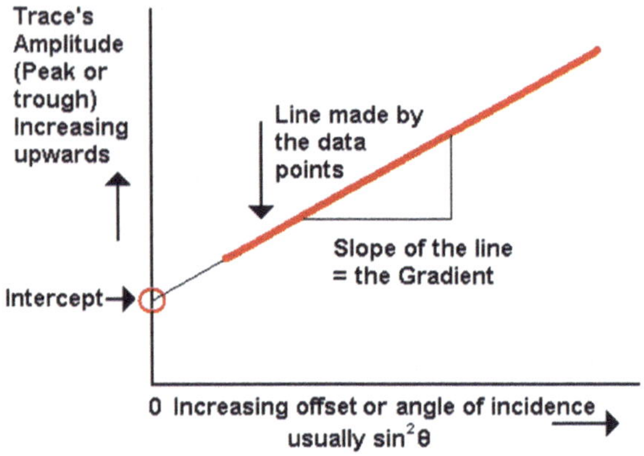

Fig. 5.11 Crossplot of $\sin^2\theta$ with intercept

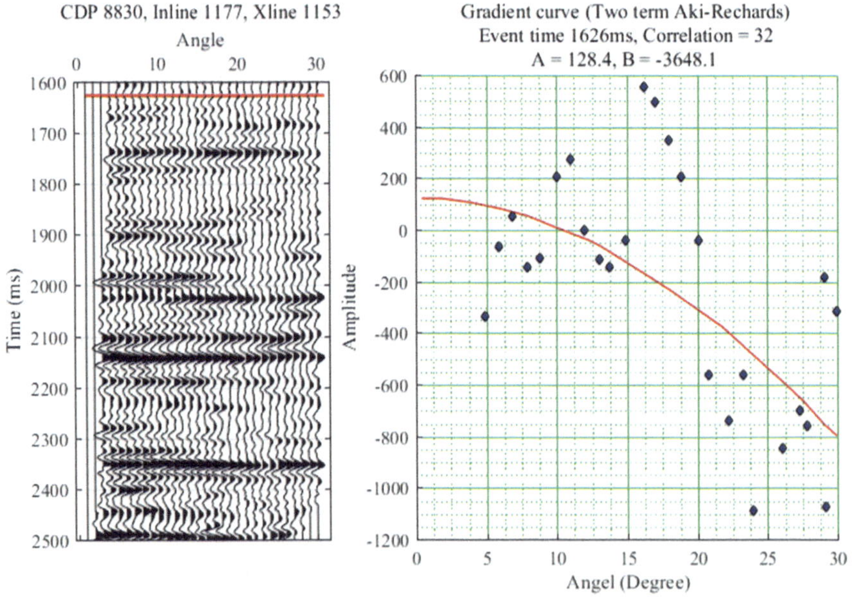

Fig. 5.12 Analysis of AVO anomalies at around 1626 ms. The red line on the seismic display shows the time location at which the amplitudes have been extracted. Those amplitudes are plotted as squares on the right-hand graph. The curve which has fit through the picks is a plot of Aki-Richards two-term equation

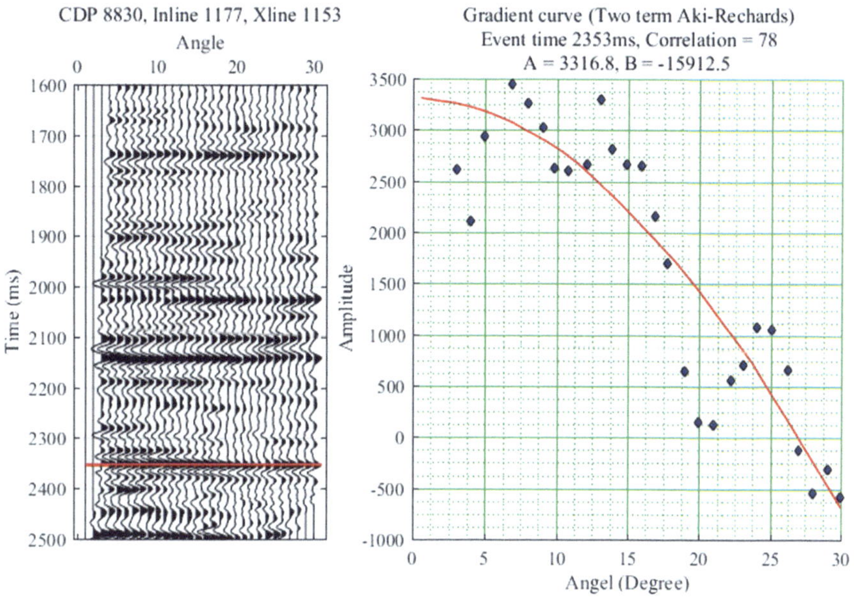

Fig. 5.13 Analysis of AVO anomalies at around 2353 ms. The red line on the seismic display shows the time location at which the amplitudes have been extracted. Those amplitudes are plotted as red squares on the right-hand graph. The curve which has fit through the picks is a plot of Aki-Richards two-term equation

Figures 5.13, 5.14 and 5.15 show the curves of the amplitude variations with incidence angles for several CMP gather selected from the dataset. These amplitude curves are utilized to estimate the intercept (*A*) and gradient (*B*) values. We would like to analyze the AVO anomaly at around 1626, 2353 and 2365 ms. The anomaly plot is shown in Fig. 5.13. The red line on the seismic display shows the time location at which the amplitudes have been extracted. Those amplitudes are plotted as red squares on the right-hand graph. The curve which has fit through the picks is a plot of Aki-Richards two-term equation. We can confirm this by the information at the top of the graph.

Notice that we are seeing a classic class 2p AVO anomaly with amplitudes increasing for both the trough at the top of the sand (red) and the peak at the base of the sand (green). Notice also that the fit of the AVO curves is good. Mathematically, this is expressed by the normalized correlation between the picked amplitudes and the curves, printed at the top of the graph. The analysis for time 2353 ms and 2365 ms are shown in Figs. 5.14, and 5.15 respectively. The graph shows class 2p anomaly for 2353 ms time and class 4 anomaly for 2365 ms time respectively.

b. **AVO attributes volume**

AVO attributes volumes such as intercept (A), gradient (B) and their proportions such as AVO product (AB), Poisson coefficient (A + B), shear reflectivity (A-B)

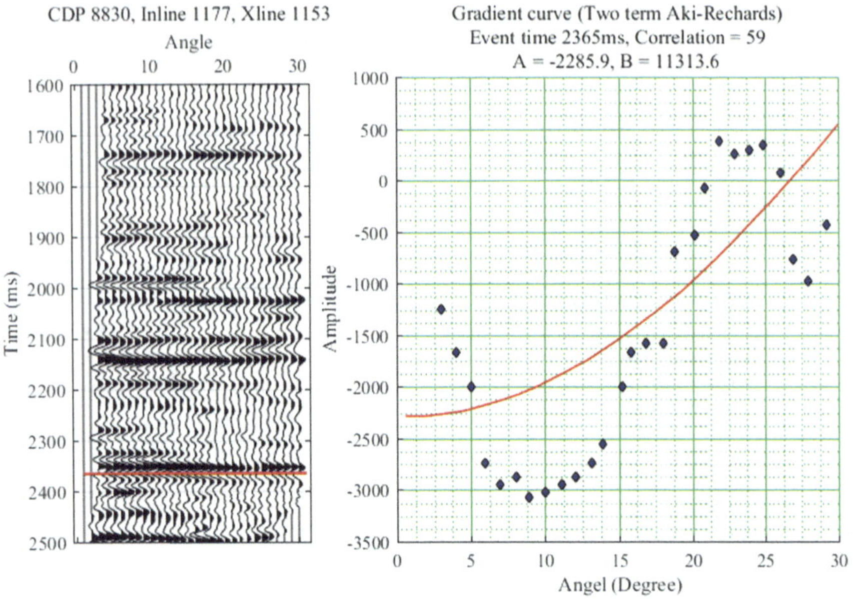

Fig. 5.14 Analysis of AVO anomalies at around 2363 ms. The red line on the seismic display shows the time location at which the amplitudes have been extracted. Those amplitudes are plotted as red squares on the right-hand graph. The curve which has fit through the picks is a plot of Aki-Richards two-term equation

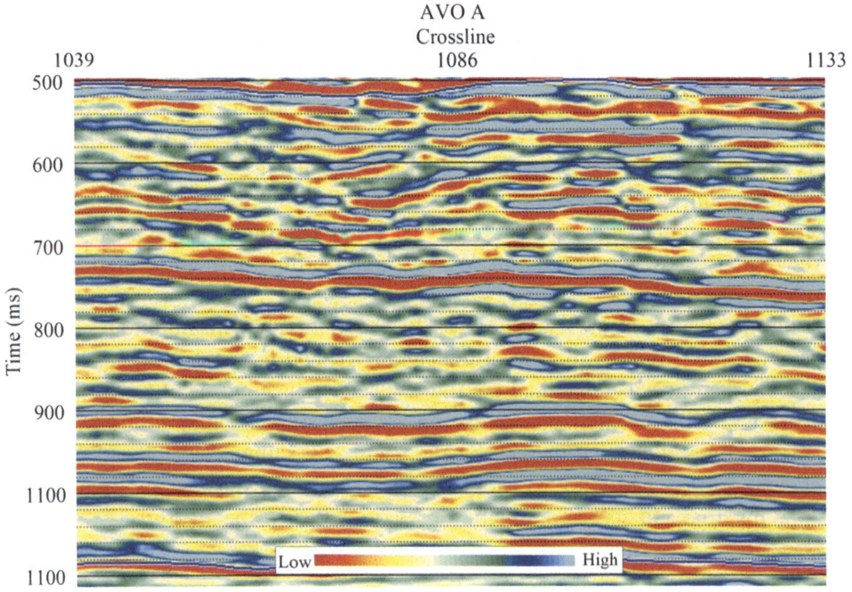

Fig. 5.15 Cross-section of intercept (inline 1163) estimated from Aki-Richards two-term equation

and fluid coefficient (FF) can be generated straight from pre-stack seismic NMO-corrected CMP-sorted data. Now we use AVO Gradient Analysis to examine the AVO anomaly, we will extend the calculation to the whole sample to see the allocation of AVO anomalies. The method of the AVO Attribute utilizes the Aki-Richards equation of two or three terms to obtain AVO attributes from the seismic data. The attributes are based on intercept, gradient and curvature combinations as described by the equation of Aki-Richards. The angle gathers volume is used as input. Different combinations of AVO attributes can be generated as per requirement. Here, two basic types of attributes namely AVO A and AVO B are generated.

The Aki-Richards approximation for two terms only is used because we only have incident angles less than 30°. We need high angle data, usually exceeding 45°, to reliably obtain three terms. The calculated attributes from the Penobscot field, Canada are shown in Figs. 5.15 and 5.16 for AVO A and AVO B respectively. One can notice the variation of intercept and gradient attributes and hence can utilize to detect the hydrocarbon zone. Although, the present datasets did not have the presence of any major hydrocarbon reservoir.

Cross-plotting of the intercept (A) against gradient (B) is an efficient interpretation technique for identification of AVO anomalies (Fig. 5.17). This method was developed by Castagna et al. (1998). It is based on two ideas: (1) the Rutherford and Williams (1989) AVO classification scheme described below and (2) the so-called Mudrock Line. The cross plot shows the anticipated context pattern through the origin, with anomalous occurrences in quadrants 1 and 3, corresponding with anomalies

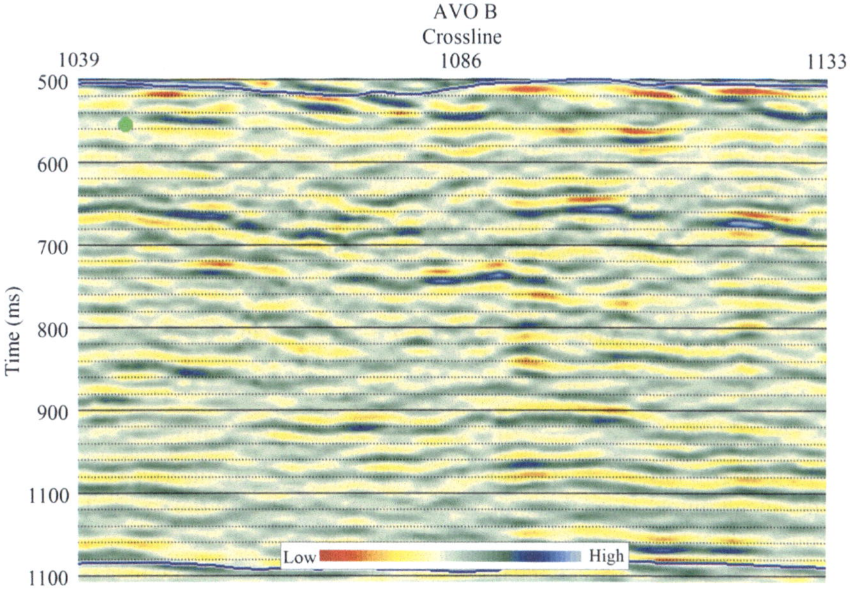

Fig. 5.16 Cross-section of the gradient (inline 1163) estimated from the Aki-Richards two-term equation

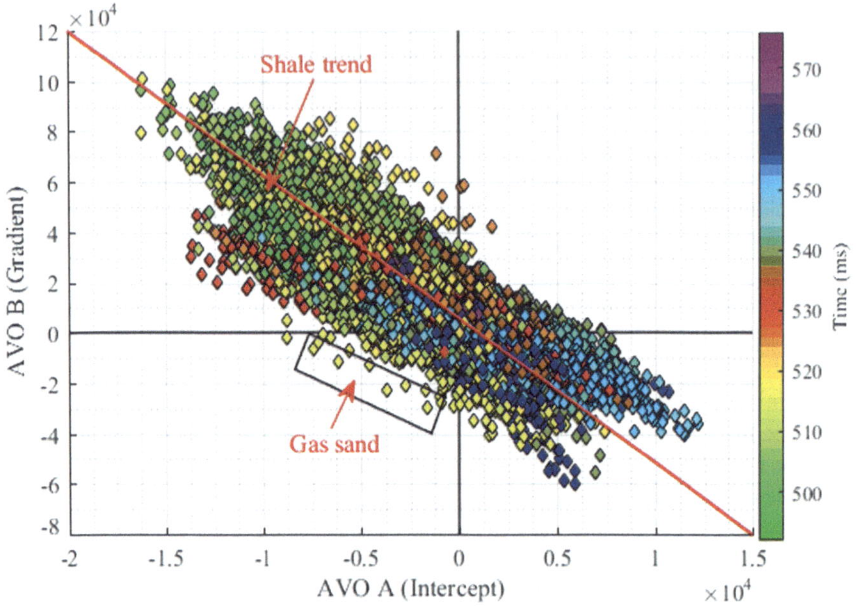

Fig. 5.17 Crossplot of intercept and gradient near to horizon 2 from Penobscot data, Canada. The Class 3 anomaly is highlighted

in class 3 AVO (Fig. 5.17). The slices of a different combination of attributes are plotted in Fig. 5.18. These slices show the nice variation of extracted attributes in the subsurface.

iv. **Shuey approximation**

Another useful simplification of the Zoeppritz equation was proposed by Shuey (1985), who decomposed the reflectivity into the normal-incidence term and correction terms principally depending on the Poisson's ratio and density variations across the boundary. The Shuey approximation can be written as follows.

$$R(\theta) \approx R_0 + \left[A_0 R_0 + \frac{\Delta\sigma}{(1-\sigma)^2}\right]\sin^2\theta + \frac{1}{2}\frac{\Delta V_P}{V_P}(\tan^2\theta - \sin^2\theta) \tag{5.42}$$

In Eq. 5.42, $R_0 = \frac{1}{2}\left[\frac{\Delta V_P}{V_P} + \frac{\Delta\rho}{\rho}\right]$, $A_0 = B - 2(1+B)\frac{1-2\sigma}{1-\sigma}$ and $B = \frac{\frac{\Delta V_P}{V_P}}{\frac{\Delta V_P}{V_P} + \frac{\Delta\rho}{\rho}}$.

The first term in Eq. 5.42 describes the amplitude at $\theta = 0$, the second term represents an amplitude correction at intermediate angles, and the third term describes the amplitude at wide angles. For a rock sample under unidirectional pressure, the Poisson's ratio σ is the ratio of the transverse expansion to the longitudinal compression, or the ratio of shear strain to principal strain (Yilmaz 2001). For isotropic rock, σ can be expressed through the ratio of the P- and S-wave velocity as follows.

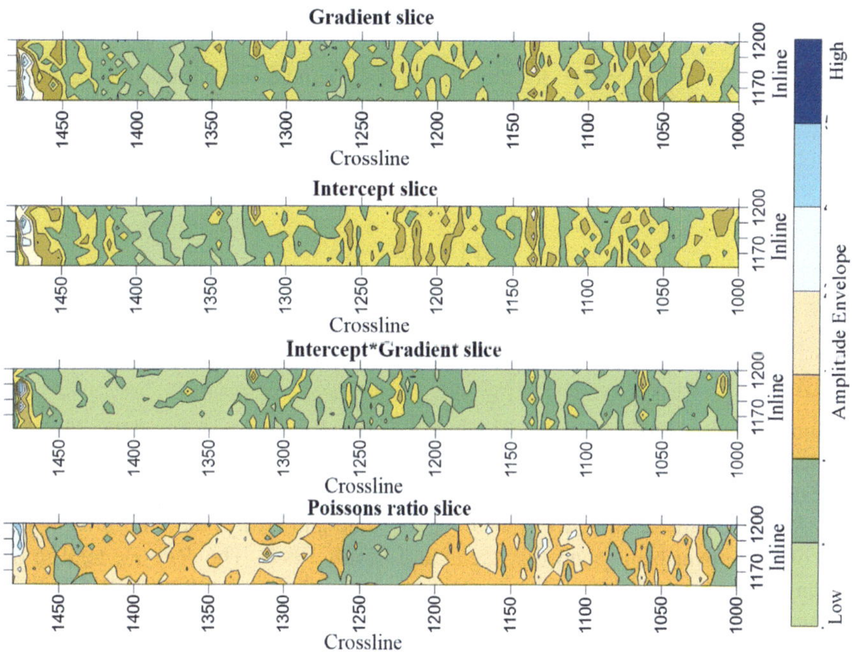

Fig. 5.18 Depicts slices from gradient (B), Intercept (A), A*B and Poisson's ratio (From top to bottom)

$$\sigma = \frac{\left(\frac{V_P}{V_S}\right)^2 - 2}{2\left(\frac{V_P}{V_S}\right)^2 - 2} \tag{5.43}$$

Thus, the Poisson's ratio increases when V_P/V_S increases, and vice versa, and therefore this ratio is typically low for gas reservoirs (Ma and Morozov 2010). It typically equals to 0.1, for gas sands. Changes in gas or fluid saturation change the Poisson's ratio significantly because of the changes in rock bulk modulus and consequently the P-wave velocity. At the same time, the shear modulus changes only slightly and therefore fluid saturation has little effect on the S-wave velocity (Gassmann 1951). Further, an increase in fluid saturation decreases the P-wave reflectivity and hence the Poisson's ratio.

v. **Fatti approximation**

An alternate form of the Aki and Richards's equation was given by Fatti et al. (1994). This expression can be written as follows.

$$R(\theta) = c_1 R_P + c_2 R_S + c_3 R_D \tag{5.44}$$

In Eq. 5.44, $c_1 = 1 + \tan^2\theta$, $c_2 = -8\gamma\sin^2\theta$, $\gamma = \left(\frac{V_P}{V_S}\right)^2$, and $c_3 = -\frac{1}{2}\tan^2\theta + 2\gamma^2\sin^2\theta$ where R_P and R_S are the P- and S-wave reflectivities which can be expressed as follows.

$$R_P = \frac{1}{2}\left[\frac{\Delta V_P}{V_P} + \frac{\Delta\rho}{\rho}\right] \tag{5.45}$$

$$R_S = \frac{1}{2}\left[\frac{\Delta V_S}{V_S} + \frac{\Delta\rho}{\rho}\right] \tag{5.46}$$

$$R_D = \frac{\Delta\rho}{\rho} \tag{5.47}$$

Equation (5.44) allows us to calculate R_P and R_P from seismic data. The difference between the P-wave and S-wave reflectivities, $(R_P - R_S)$, can be used as an indicator differentiating the shale over brine-sand and shale over gas-sand cases. $R_P - R_S$ values are negative for shale over gas-sand and always more negative in the case of shale over brine-sand (Castagna and Smith 1994). The $R_P - R_S$ tend to be constant and near zero for non-pay reservoirs. The reflectivities R_P and R_S can also be transformed into two new attributes i.e. the Fluid Factor (FF) and Lambda-Mu-Rho (LMR). Figures 5.19, 5.20 and 5.21 shows the variation of R_P, R_S and cross the plot between them. Figure 5.22 shows cross-section of $R_P - R_S$ respectively. A

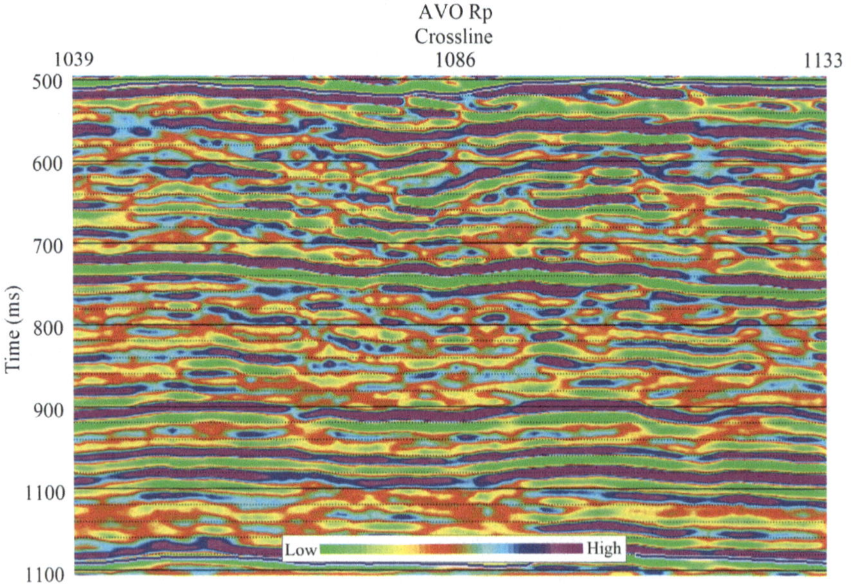

Fig. 5.19 Cross-section (inline 1165) of inverted R_P attributes estimated from Fatti et al. (1994) two-term equation

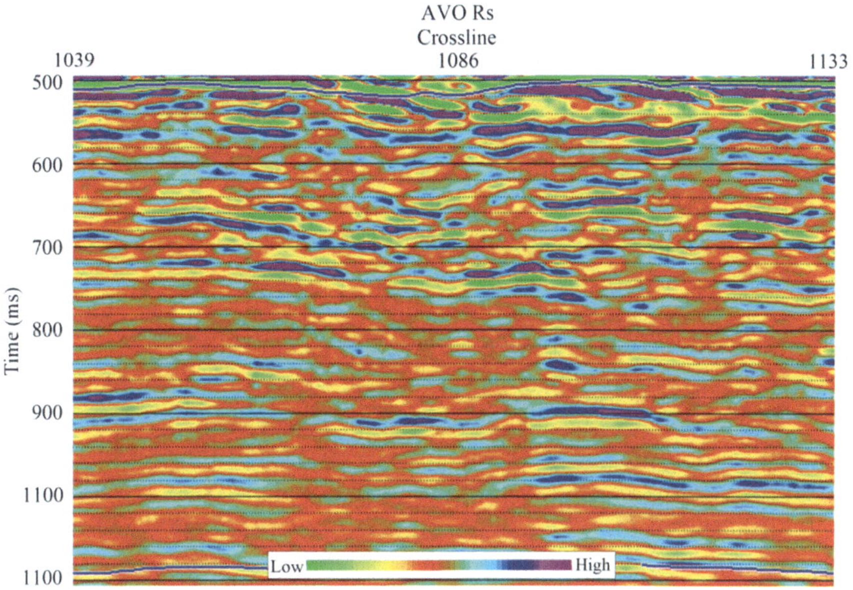

Fig. 5.20 Cross-section (inline 1165) of inverted R_S attributes estimated from Fatti et al. (1994) two-term equation

Fig. 5.21 Crossplot of R_P versus R_S and the anomalous zone is highlighted

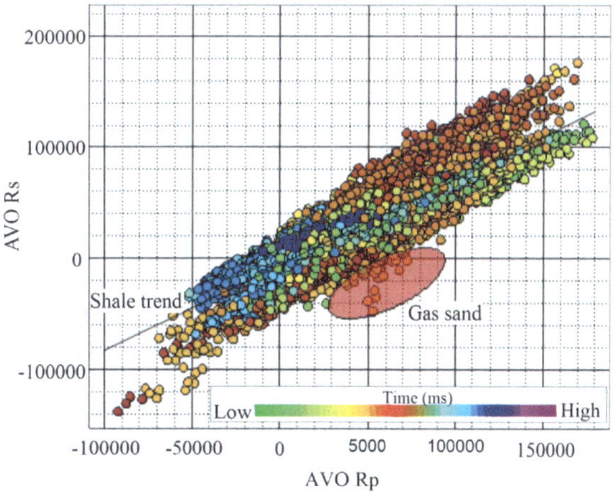

small anomaly zone can be seen from the cross plot (Fig. 5.21).

In a clastic sedimentary sequence, the Fluid Factor is defined so that it is high-amplitude for reflectors that lie far from the mudrock line and low-amplitude for all reflectors on the mudrock line. The equation defining the FF, according to Fatti et al. (1994) can be written as follows.

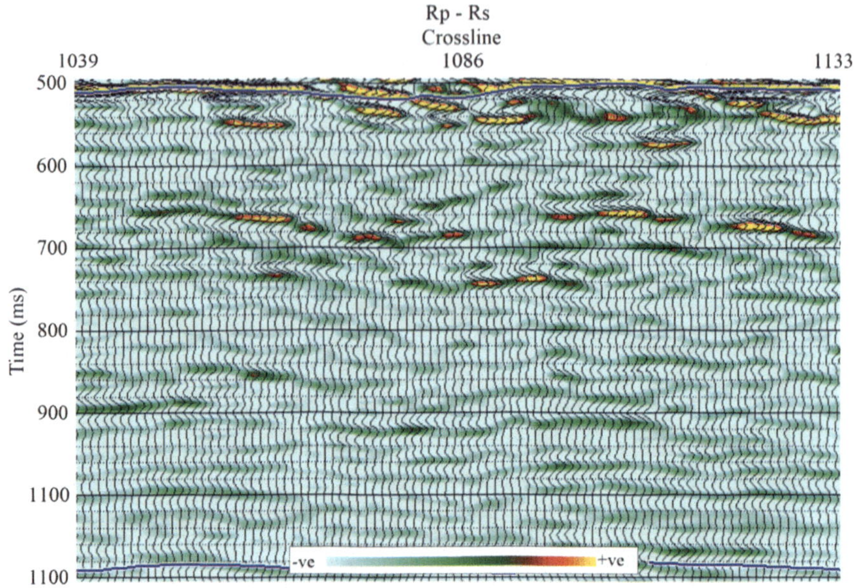

Fig. 5.22 Cross-section of $R_P - R_S$ at inline 1165. The anomalous zone is clearly visible

$$\Delta F = \frac{\Delta V_P}{V_P} - 1.16\frac{V_S}{V_P}\frac{\Delta V_S}{V_S} \tag{5.48}$$

This can also be written as follows.

$$\Delta F = R_P - 1.16\frac{V_S}{V_P}R_S \tag{5.49}$$

The FF equals zero when layers both above and below the reflecting boundary lie on the mudrock line, such as shale over brine sand. By contrast, the FF is nonzero when one of the layers lies on the mudrock line and the other one lies away from it (Fatti et al. 1994). In cases of gas sands, the FF will be non-zero at both the top and bottom of gas.

The Lambda-Mu-Rho attributes (LMR) are defined so that the Lame's elastic parameters λ and μ are combined with density ρ in the form of $\lambda\rho$ and $\mu\rho$, as was first proposed by Goodway et al. (1997). Pre-stack seismic CMP gathers are inverted to extract the *P*-impedance and *S*-impedance, and from these impedances, the $\lambda\rho$ and $\mu\rho$ products are extracted. Starting from the equations for wave velocities, we have the followings relation.

$$V_P = \sqrt{\frac{\lambda + 2\mu}{\rho}} \tag{5.50}$$

$$V_S = \sqrt{\frac{2\mu}{\rho}} \tag{5.51}$$

We have

$$\mu\rho = (V_S\rho)^2 = Z_S^2 \tag{5.52}$$

$$(V_P\rho)^2 = Z_P^2 = (\lambda + 2\mu)\rho \tag{5.53}$$

And therefore

$$\lambda\rho = Z_P^2 - 2Z_S^2 \tag{5.54}$$

The λ parameter, or incompressibility, is sensitive to pore fluid, whereas the μ factor, or rigidity, is sensitive to the rock matrix.

References

Aki K, Richards PG (1980) Quantitative seismology. W.H. Freeman and Company, San Francisco, CA

Aki K, Richards PG (2002) Quantitative seismology, 2nd edn. University Science Books, Sausalito, CA

Avseth P, Bachrach R (2005) Seismic properties of unconsolidated sands: tangential stiffness, Vp/Vs ratios and diagenesis. In: SEG Technical Program Expanded Abstracts, Society of Exploration Geophysicists, pp 1473–1476

Backus MM, Chen RL (1975) Flat spot exploration. Geophys Prospect 23(3):533–577

Bacon M, Simm R, Redshaw T (2007) 3-D seismic interpretation. Cambridge University Press, Cambridge

Batzle M, Wang Z (1992) Seismic properties of pore fluids. Geophysics 57(11):1396–1408

Brown AR (2012) Dim spots: opportunity for future hydrocarbon discoveries? Lead Edge 31(6):682–683

Castagna JP, Backus MM (eds) (1993) Offset-dependent reflectivity—theory and practice of AVO analysis. Soc Explor Geophys

Castagna JP, Smith SW (1994) Comparison of AVO indicators: a modeling study. Geophysics 59(12):1849–1855

Castagna JP, Swan HW (1997) Principles of AVO cross plotting. Lead Edge 16(4):337–344

Castagna JP, Swan HW, Foster DJ (1998) Framework for AVO gradient and intercept interpretation. Geophysics 63(3):948–956

Chen Q, Sidney S (1997) Seismic attribute technology for reservoir forecasting and monitoring. Lead Edge 16(5):445–448

Chiburis E, Leaney S, Skidmore C, Franck C, McHugo S (1993) Hydrocarbon detection with AVO. Oilfield Rev 5:1–3

Coulombe AC (1993) Amplitude versus offset analysis using vertical seismic profiling and well log data. The crewes *project*, consortium for research in elastic wave exploration seismology, 167p

Dey-Sarkar SK, Foster DJ, Smith SW, Swan HW, Atlantic Richfield Co. (1995) Seismic data hydrocarbon indicator. U.S. Patent 5: 440,525

Dufour J, Goodway B, Shook I, Edmunds A et al (1998) AVO analysis to extract rock parameters on the Blackfoot 3c-3d seismic data. In: 68th Annual International SEG Mtg, pp 174–177

Dufour J, Squires J, Goodway WN, Edmunds A, Shook I (2002) Integrated geological and geophysical interpretation case study, and lame rock parameter extractions using AVO analysis on the Blackfoot 3C-3D seismic data, southern Alberta, Canada. Geophysics 67(1):27–37

Fatti JL, Smith GC, Vail PJ, Strauss PJ, Levitt PR (1994) Detection of gas in sandstone reservoirs using AVO analysis: a 3-d seismic case history using the geo-stack technique. Geophysics 59(9):1362–1376

Gardner GHF, Gardner LW, Gregory AR (1974) Formation velocity and density —the diagnostic basics for stratigraphic traps. Geophysics 39:770–780

Gassmann F (1951) Elastic waves through a packing of spheres. Geophysics 16(4):673–685

Goodway B, Chen T, Downton J (1997) Improved AVO fluid detection and lithology discrimination using Lamé petrophysical parameters; "$\lambda\rho$", "$\mu\rho$", & "λ/μ fluid stack", from P and S inversions. In: SEG Technical Program Expanded Abstracts, Society of Exploration Geophysicists, pp 183–186

Hammond AL (1974) Bright spot: better seismological indicators of gas and oil. Science 185(4150):515–517

Han DH, Batzle ML (2004) Gassmann's equation and fluid-saturation effects on seismic velocities. Geophysics 69(2):398–405

Hashin Z, Shtrikman S (1963) A variational approach to the theory of the elastic behavior of multiphase materials. J Mech Phys Solids 11(2):127–140

Hill R (1952) The elastic behavior of a crystalline aggregate. Proc Phys Soc. Sect A 65(5):349

Inyang CB (2009) AVO analysis and impedance inversion for fluid prediction in Hoover Field, Gulf of Mexico, Doctoral dissertation, University of Houston

Keys RG (1989) Polarity reversals in reflections from layered media. Geophysics 54(7):900–905

Knott CG (1899) Reflexion and refraction of elastic waves, with seismological applications. Lond Edinburgh Dublin Philosop Maga J Sci 48(290):64–97

Koefoed O (1955) On the effect of Poisson's ratios of rock strata on the reflection coefficients of plane waves. Geophys Prospect 3(4):381–387

Koefoed O (1962) Reflection and transmission coefficients for plane longitudinal incident waves. Geophys Prospect 10(3):304–351

Kumar D (2006) A tutorial on Gassmann fluid substitution: formulation, algorithm and Matlab code. Matrix 2:1

Larsen JA, Margrave GF, Lu HX (1999) AVO analysis by simultaneous PP and PS weighted stacking applied to 3C-3D seismic data. In: SEG Technical Progress Expanded Abstract, Society of Exploration Geophysicists, pp 721–724

Li XY, Mueller MC (1997) Case studies of multicomponent seismic data for fracture characterization: Austin Chalk examples. In: Carbonate seismology, Society of Exploration Geophysicists, pp 337–372

Li Y, Downton J, Xu Y (2007) Practical aspects of AVO modeling. Lead Edge 26(3):295–311

Luping G (2008) Pre-stack inversion and direct hydrocarbon indicator. Geophys Prospect Petroleum, 3

Ma J, Morozov I (2010) AVO modeling of pressure-saturation effects in Weyburn CO_2 sequestration. Lead Edge 29(2):178–183

Maurya SP, Singh NP (2019) Seismic modelling of CO_2 fluid substitution in a sandstone reservoir: a case study from Alberta, Canada. J Earth Syst Sci 128(8):236

Mavko G, Mukerji T (1998) Bounds on low-frequency seismic velocities in partially saturated rocks. Geophysics 63(3):918–924

Mocnik A (2011) Processing and analysis of seismic reflection data for hydrocarbon exploration in the plio-quaternary marine sediments, Doctoral thesis, University of Trieste

Muskat M, Meres MW (1940) Reflection and transmission coefficients for plane waves in elastic media. Geophysics 5(2):115–148

Ostrander W (1984) Plane-wave reflection coefficients for gas sands at nonnormal angles of incidence. Geophysics 49(10):1637–1648
Pendrel J, Dickson T (2003) Simultaneous AVO Inversion to P Impedance and Vp/Vs. In: CSEG Annual Meeting, Expanded Abstract
Ramos AC, Davis TL (1997) 3-D AVO analysis and modeling applied to fracture detection in coalbed methane reservoirs. Geophysics 62(6):1683–1695
Ross CP, Kinman DL (1995) Nonbright-spot AVO: two examples. Geophysics 60(5):1398–1408
Russell B (1988) Introduction to seismic inversion methods. The SEG Course Notes, Series 2
Russell H (1999) Theory of the STRATA Program. Hampson-Russell, CGG Veritas
Rutherford SR, Williams RH (1989) Amplitude-versus-offset variations in gas sands. Geophysics 54(6):680–688
Sheriff, R.E., 1973. Encyclopedic dictionary of exploration geophysics: Tulsa. *Society of Exploration Geophysicists*
Shuey R (1985) A simplification of the zoeppritz equations. Geophysics 50(4):609–614
Simmons JL, Backus MM (1996) Waveform-based AVO inversion and AVO prediction-error. Geophysics 61(6):1575–1588
Smith GC, Gidlow PM (1987) Weighted stacking for rock property estimation and detection of gas. Geophys Prospect 35(9):993–1014
Smith GC, Sutherland RA (1996) The fluid factor as an AVO indicator. Geophysics 61(5):1425–1428
Smith TM, Sondergeld CH, Rai CS (2003) Gassmann fluid substitutions: A tutorial. Geophysics 68(2):430–440
Swan HW, Castagna JP, Backus MM (1993) Properties of direct AVO hydrocarbon indicators. Offset dependent reflectivity-theory and practice of AVO anomalies. Invest Geophys 8:78–92
Tai S, Puryear C, Castagna JP (2009) Local frequency as a direct hydrocarbon indicator. In: SEG Technical Program Expanded Abstracts, Society of Exploration Geophysicists, pp 2160–2164
Verm R, Hilterman F (1995) Lithology color-coded seismic sections: the calibration of AVO cross plotting to rock properties. Lead Edge 14(8):847–853
Wang Z (2001) Fundamentals of seismic rock physics. Geophysics 66(2):398–412
White RS (1977) Seismic bright spots in the Gulf of Oman. Earth Planet Sci Lett 37(1):29–37
Wiggins R, Kenny GS, McClure CD (1983) A method for determining and displaying the shear-velocity reflectivities of a geologic formation. European patent application, 113944
Xu Y, Bancroft JC (1997) Joint AVO analysis of PP and PS seismic data. The CREWES Project Research Report, 9
Yilmaz O (2001) Seismic data analysis, vol 1. Society of Exploration Geophysicists, Tulsa, OK
Zhu X, McMechan GA (1990) Direct estimation of the bulk modulus of the frame in a fluid-saturated elastic medium by Biot theory. In: SEG Technical Program Expanded Abstracts, Society of Exploration Geophysicists, pp 787–790
Zoeppritz K (1919) On the reflection and propagation of seismic waves. Gottinger Nachrichten 1(5):66–84

Chapter 6
Optimization Methods for Nonlinear Problems

Abstract The optimization method is a process of getting the maximum or minimum of the inverse problem. This concept is used in this chapter to find the solution to the geophysical problems. Nowadays these techniques are used frequently in the exploration field to estimate subsurface information using seismic and well log data. Although these methods are used in almost all branches of geophysics but here the explanation is focused in the field of exploration. Initially, some local optimization methods like Steepest Descent Method, Conjugate Gradient Method, Newton Methods are discussed. Thereafter, two types of global optimization methods namely Genetic algorithm and Simulated Annealing are discussed along with synthetic as well as real data examples.

6.1 Introduction

An optimization method is a process of getting optimal solutions to the problem under some given circumstances. It is a process of finding the maximum or minimum of the inverse problem (Gill et al. 1981). The objective of any optimization technique is to discover with small computational effort an ideal or near-optimal solution to the problem. An optimization method's effort can be measured as the time (computing time) and space (computer memory) that the method consumes. There is a trade-off between solution quality and effort for many techniques of optimization, and particularly for contemporary heuristics, as the quality of the solution rises with growing effort (Holland 1975). The optimization methods can be understood from the following examples.

Let's assume a person moving from one place (say workplace) to another place (say home), and there are many modes of transportation available to travel. The person who can travel by Taxi/Cab is one mode of transportation. He can also take a train to reach the destination and this would be their second mode of transportation. The third mode of transportation is to use the bus to reach the destination and the fourth mode of transportation is to use multiple transportations services. Now the problem arises to decide which mode of transportation is better by considering the least cost and time. In this type of problem, the optimization technique is used to

S. P. Maurya et al., *Seismic Inversion Methods: A Practical Approach*, Springer Geophysics, https://doi.org/10.1007/978-3-030-45662-7_6

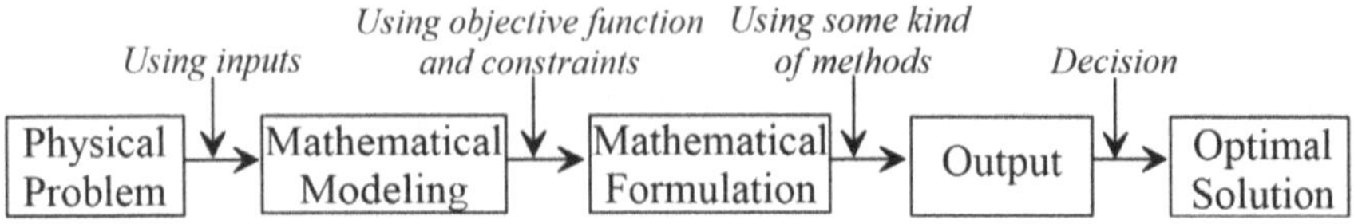

Fig. 6.1 Schematic diagram showing the process of solving nonlinear problems using an optimization technique

minimize cost and time and hence to decide better transportation service one can use to move from one place to the other place. These types of problems can be solved as follows.

First, the physical problems are transformed into mathematical modeling using the information provided in the problem. Thereafter, this mathematical modeling is transformed into mathematical formulation according to the optimization technique using objective function and constraints. Further, the solution (output) is generated by solving some kind of optimization technique. Further, the decision is made to decide either the solution is optimal or not. Graphically, it can be represented as follows (Fig. 6.1).

One of the geophysical inversion's main objectives is to discover earth models that explain geophysical findings. Thus, in many geophysical applications, the mathematics branch known as optimization has discovered important use. In this sense, geophysical inversion includes finding the ideal value of a multiple variables function.

The feature we want to minimize (or maximize) is a misfit (or fitness) function that characterizes the distinctions (or similarities) between observed and synthetic information calculated using an assumed earth model. The earth model is depicted by physical parameters that characterize rock layers' characteristics, such as compressional wave velocity, shear-wave velocity, resistivity, etc.

In the estimation of material characteristics from geophysical information, both local and global optimization techniques are used. The objective of this chapter is to explain the implementation of geophysical problems of several newly developed techniques of local and global optimizations. Although we emphasize the implementation elements of these algorithms, we explain in sufficient detail several components of the theory for readers to know the underlying basic values on which these algorithms are based.

There are many optimization techniques, and they are selected on the basis of objective function and constraints provided in the problem. For many geophysical applications, the misfit surface can be extremely complex and distinguished by various mountains and valleys as a function of the model parameters outlined by the mismatch between the expected and observed geophysical information (Jervis et al. 1996). Such a feature will, therefore, have several minima and maxima; the minimum of all minima will be called the global minimum, and all other minima will be called local minima. Note that the global minimum is one of the local minima, but the opposite is not true, and multiple minima of nearly the same depth may also be available (Sen and Stoffa 2013). They use the misfit function's local characteristics to calculate an update to the present response and search downhill direction. Thus, if the starting point is closer to one of the local minima than the global minimum, these algorithms will miss the global minimum solution. Geophysicists have been plagued by the local minimum syndrome for over a century now.

There are two broad categories available for the optimizations, the first is local optimization methods and the second is called global optimization methods. The local optimization method generally looks for a local minimum solution in the vicinity of a startup model. On the other hand, the global optimization methods have a tendency to reach the global minimum by jumping from the local minima (Sen and Stoffa 2013). These local and global optimization methods are further divided into several subparts. In the following sections, they are explained briefly. Before discussing these optimization methods, we are in a position to discuss different kinds of optimization methods that are used by both methods.

6.2 Fitness Functions

One can characterize the fitness function as continuous or discrete. When variables can take only finite numbers in the fitness function, they are called discrete problems of optimization. On the other hand, if variables in the error function can take a continuous set of real values, the continuous optimization problems arise (Dorrington and Link 2004; Sen and Stoffa 2013). We have provided five types of fitness function produced by alteration in the l_1 and l_2 norm. These fitness functions are mostly used in all geophysical problems. They are as follows.

The first fitness function we are presenting here is based on L_2 norm. This L_2 norm is applied between observed and modeled data (Moncayo et al. 2012). It can be represented as follows.

$$E = \frac{1}{n}\sqrt{\sum_{i=1}^{n}\left(S_{obji} - S_{modi}\right)^2} \tag{6.1}$$

In Eq. 6.1, E is the error between modeled and observed data, S_{obji} is observed data at ith sample, S_{modi} is modeled data at ith sample and n is the total number of sample points in the data. This method is largely used in geophysics because of fast convergence.

The second fitness equation is taken from the mean square error between modeled and observed data (Pullammanappallil and Louie 1994). The mathematical formulae can be expressed as follows.

$$E = \frac{1}{n}\sum_{i=1}^{n}\left(S_{obji} - S_{modi}\right)^2 \tag{6.2}$$

The fitness function presented in Eq. 6.2 takes a relatively larger time to converge as compared to the first fitness function.

The third fitness equation is borrowed from the Misra and Sacchi (2008) which can be expressed as follows.

$$E = \sum_{i=1}^{n} \left(\left| S_{objı} - S_{modi} \right| \right)^2 \tag{6.3}$$

This fitness equation is very time consuming and hence before using them, proper care should be given.

The fourth fitness equation presented here is based on l_1 norm. This is also called the least absolute deviation. This is calculated between observed and modeled data sets sample by sample. In order to constrain the solution, one additional term is added to the equation and this additional term is called a priori impedance model. Mathematically this fitness function can be expressed as follows.

$$E = W_1 \left(\frac{\sum_{i=1}^{n} \left| S_{obji} - S_{modi} \right|}{\sum_{i=1}^{n} \left| S_{obj} \right|} \right) + W_2 \left(\frac{\sum_{i=1}^{n} \left| Z_{obji} - Z_{modi} \right|}{\sum_{i=1}^{n} \left| Z_{obj} \right|} \right) \tag{6.4}$$

In Eq. 6.4, W_1 and W_2 are weights applied to the two terms, respectively. The weighting factor can be chosen as $W_1 = W_2 = 1$ in most cases (Ma 2002). The Z_{prii} is the prior low-frequency impedance at ith time sample and Z_{modi} is the model impedance at ith time sample. The last (fifth) fitness equation is again based on root mean square error (l_2 norm). In this equation, one additional term is added which comes from the initial impedance model. This additional term act as constraints for the solution (Ma 2002). Using this equation the non-uniqueness of the solution can be reduced to some level. The expression can be written as follows.

$$E = \frac{1}{n} \sqrt{\sum_{l=1}^{n} \left(S_{obji} - S_{modi} \right)^2} + \frac{1}{n} \sqrt{\sum_{i=1}^{n} \left(Z_{prii} - Z_{modi} \right)^2} \tag{6.5}$$

In Eq. 6.5, the Z_{prii} is the prior low-frequency impedance at ith time sample and Z_{modi} is the modeled impedance at ith time sample. The convergence of this fitness function (Eq. 6.5) is very fast.

6.3 Local Optimization Methods

The local optimization algorithms move from solution to solution in the search space of candidate solutions by applying local changes, until a solution deemed optimal is found. Most of the local optimization methods are solved in an iterative manner, and the objective of these algorithms is to move towards the solution in each iteration and hence guarantee a reduction in the value of the objective function (Bremermann 1958, 1962). The local optimization methods get stuck at the local minimum in the vicinity of the initial model and therefore, the choice of the initial model is of utmost importance. If a choice of initial guess model will be made away from the global solution then, the algorithm will be trapped in a local minimum. The basic working principle of local optimization methods is presented in Fig. 6.2.

The local optimization methods start with a given startup/initial model. Thereafter, objective function is employed to estimate error and if this error is small enough, then the initial guess model will be treated as a final solution. If the error is not small enough, the algorithm calculates a search address using a local property of the objective function, which determines an update or increment of the current one model (Hejazi et al. 2013). There are many local optimization methods available but some of them are as follows.

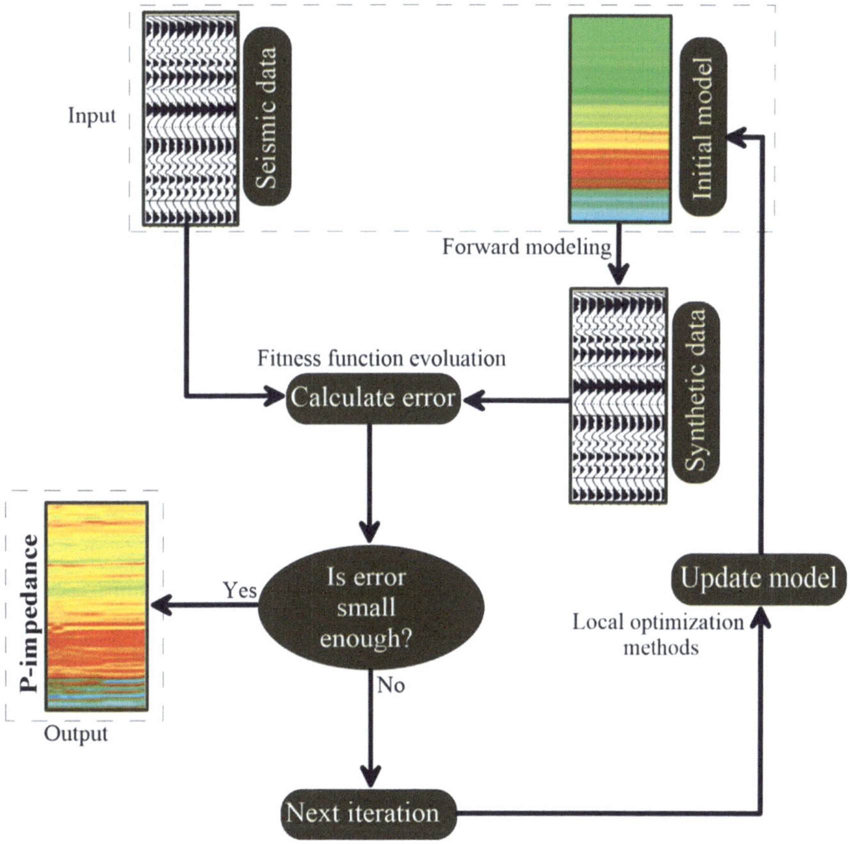

Fig. 6.2 Flowchart of local optimization methods

1. Steepest Descent Method
2. Conjugate Gradient Method
3. Newton Methods.

6.3.1 Steepest Descent Method

The steepest descent method is an algorithm used to find the nearest local minimum of an inverse problem which presupposes that the gradient of the inverse problem can be computed. This method is also called the gradient descent method (Sen and Stoffa 2013; Deift and Zhou 1993). Mathematically it can be expressed as follows.

Let's a function $f : \mathbb{R}^n \to \mathbb{R}$ is such that the function f is differentiable at some point x_0, and then the direction of steepest descent is the vector which can be estimated as $-\nabla f(x_0)$. Now, consider the function that can be written as follows.

$$\varphi(t) = f(x_0 + tu), \tag{6.6}$$

In Eq. 6.6, u is a unit vector; that is, $\|u\| = 1$. Now, partial differentiation is applied to Eq. 6.6, and we have the following equation.

$$\varphi'(t) = \frac{\partial f}{\partial x_1}\frac{\partial x_1}{\partial t} + \ldots + \frac{\partial f}{\partial x_n}\frac{\partial x_n}{\partial t} \tag{6.7}$$

$$\varphi'(t) = \frac{\partial f}{\partial x_1}u_1 + \ldots + \frac{\partial f}{\partial x_n}u_n$$

$$\varphi'(t) = \nabla f(x_0 + tu).u,$$

And therefore

$$\varphi'(0) = \nabla f(x_0).u = \nabla f(x_0)\cos(\theta), \tag{6.8}$$

where θ is the angle between $\nabla f(x_0)$ and u. It follows that $\varphi'(0)$ is minimized when $\theta = \pi$, which yields the following equation.

$$u = \frac{\nabla f(x_0)}{\nabla f(x_0)}, \varphi'(0) = -\nabla f(x_0) \tag{6.9}$$

From Eq. 6.9, we are noticed that the problem is reduced to the single variable from the several variables. Now, we want to find the minimum of $\varphi(t)$ for $t > 0$,

$$\varphi_0(t) = f(x_0 - t\nabla f(x_0)) \tag{6.10}$$

Further, after finding the minimizer t_0, we can set

$$x_1 = x_0 - t_0 \nabla f(x_0) \tag{6.11}$$

and continue the process, by searching from x_1 in the direction of $-\nabla f(x_0)$ to obtain x_2 by minimizing $\varphi_1(t) = f(x_1 - t\nabla f(x_1))$, and so on.

This is the Method of Steepest descent, given an initial guess x_0, the method computes a sequence of iterates $\{x_k\}$, where

$$x_{k+1} = x_k - t_k \nabla f(x_k), k = 0, 1, 2, \ldots, \tag{6.12}$$

where $t_k > 0$ minimizes the function

$$\varphi_k(t) = f(X_k - t\nabla f(x_k)) \tag{6.13}$$

6.3.2 *Conjugate Gradient Method*

The conjugate gradient method is again local optimization methods used to find the optimum solution to the inverse problems. It is a mathematical technique used to solve the linear and non-linear inverse problem in an iterative manner. Although, the method can also be used as a direct search technique and hence produce a numerical solution to the problem (Polyak 1969; Sen and Stoffa 2013). It is a very useful algorithm to solve large dimensional linear and non-linear inverse problems. The method is very popular because of the very fast convergence rate and also need less space to store computed data (Shewchuk 1994). The inverse problems are solved by using the conjugate gradient method in the following way.

First, compute synthetic data $d_n = g(m_n)$ and then compute the derivative of data with respect to the current model given by $G_n(m_n)$. Thereafter, compute data residual $\Delta d = d_n - d_{obs}$, and model residual $\Delta m_n = m_n - m_{pr}$. Further, compute the regularized objective function as follows.

$$E(m_n) = \frac{1}{2}\left[\Delta d_n^T \Delta d_n + \Delta m_n^T \Delta m_n\right] \tag{6.14}$$

Thereafter, compute the direction of steepest ascent. This can be achieved as follows.

$$\gamma_n = \left[G_n^T \Delta d_n + \Delta m_n\right] \tag{6.15}$$

Further, compute the conjugate direction $\varphi_n = \gamma_n + \sigma_n \varphi_{n-1}$, such that $\varphi_0 = \gamma_0$ and

$$\sigma_n = \frac{(\gamma_n - \gamma_{n-1})^T \gamma_n}{\gamma_{n-1}^T \gamma_{n-1}} \tag{6.16}$$

In last, compute optimum step length μ_n using a linear search and update model in the following way.

$$m_{n+1} = m_n - \mu_n \varphi_n \tag{6.17}$$

These steps repeated until one finds an optimum solution to the inverse problem.

6.3.3 Newton Method

Newton method is again an iterative local optimization method used to find roots of an inverse problem which is differentiable. The method is applied to the derivative f' of a twice twice-differentiable function f to find the roots of the derivative (Dennis and Moré 1977; Sen and Stoffa 2013). The basic process of the Newton method is to construct a sequence x_n from an initial guess model x_0 which have tendency to converge towards some point x^* such that $f(x^*) = 0$.

The second-order Taylor expansion $f_T(x)$ of f around x_n gives the following expression.

$$f_T(x) = f_T(x_n + \Delta x) \approx f(x_n) + f'(x_n)\Delta x + \frac{1}{2} f''(x_n)\Delta x^2 \tag{6.18}$$

Now, we want to find Δx such that $x_n + \Delta x$ is a stationary point. By differentiation of Eq. 6.18, we will have the following equation.

$$0 = \frac{d}{d\Delta x}\left(f(x_n) + f'(x_n)\Delta x + \frac{1}{2} f''(x_n)\Delta x^2 \right) = f'(x_n) + f''(x_n)\Delta x \tag{6.19}$$

$$\Delta x = -\frac{f'(x_n)}{f''(x_n)} \tag{6.20}$$

Equation 6.20 is the solution to the equation. In this solution, it is hoped that $x_{n+1} = x_n + \Delta x = x_n - f'(x_n)/f''(x_n)$ will move closer to a stationary point x^* (Kelley 1999).

Many inverse geophysical problems are nonlinear optimization problems. Local optimization techniques, e.g. generalized linear inversion, steepest descent, etc., therefore do not provide a satisfactory solution as they usually converge to a local minimum, based on the starting model's selection (Broyden 1967). Therefore, global methods of optimization such as genetic algorithms or simulated annealing are appropriate to solve these issues.

6.4 Global Optimization Methods

The objective of the global optimization method is to find the optimum solution globally of the inverse problem in the presence of many local solutions. Recently, global optimization methods are routinely used to solve the geophysical inverse problem because of the availability of advanced computing systems and large memory. The global optimization methods attempt to find the global optimum solution of the misfit function rather than finding a local optimum solutions like local optimization methods (Sen et al. 1995; Kelley 1999). A solution is called global optimal if there are no other feasible solutions available with better objective function values. On the other hand, a solution is termed as a locally optimal solution if there is no other solution with better objective function value is available in the vicinity (Kelley 1999). These objective functions play a very important role to optimize the problem.

The main advantages of using global optimization methods over local optimization methods are not dependent on the initial guess model. There are many global optimization techniques are available but some important algorithm which is regularly used in geophysical problems are as follows.

1. Genetic algorithm (GA).
2. Simulated annealing (SA).

These methods are discussed briefly in the following sections. These methods are largely used in geophysical problems hence they are applied, first, synthetic data sets and then to the real data sets to estimate subsurface rock property. Figure 6.3 depicts flowchart of global optimization methods.

6.4.1 Genetic Algorithm (GA)

Holland (1975) developed the genetic algorithm. The GA is based on analogies with biological evolution procedures. The genetic algorithm solves constrained and unconstrained problems of optimization based on natural selection, the process that drives biological development (Du and MacGregor 2010; Maurya et al. 2017). The fundamental mechanism behind GA is that to get the optimum solution, it constantly modifies a population's individual solution. The genetic algorithm randomly chooses each individual solution, which is referred to as parents and used to create a fresh solution called children (Mallick and Subhashis 1999). In many branches of science, the genetic algorithm is used to optimize discontinuous, non-differentiable, stochastic, or highly nonlinear problems by minimizing errors between observed and modeled data. There are three fundamental genetic algorithm operations that are conducted from the present population to generate the new population. They are selection, crossover, and mutation. The selection is used to select the individual's solution of the population (parents) however the crossover is used to combine two parents and hence to form children. The last genetic operator mutation is used to

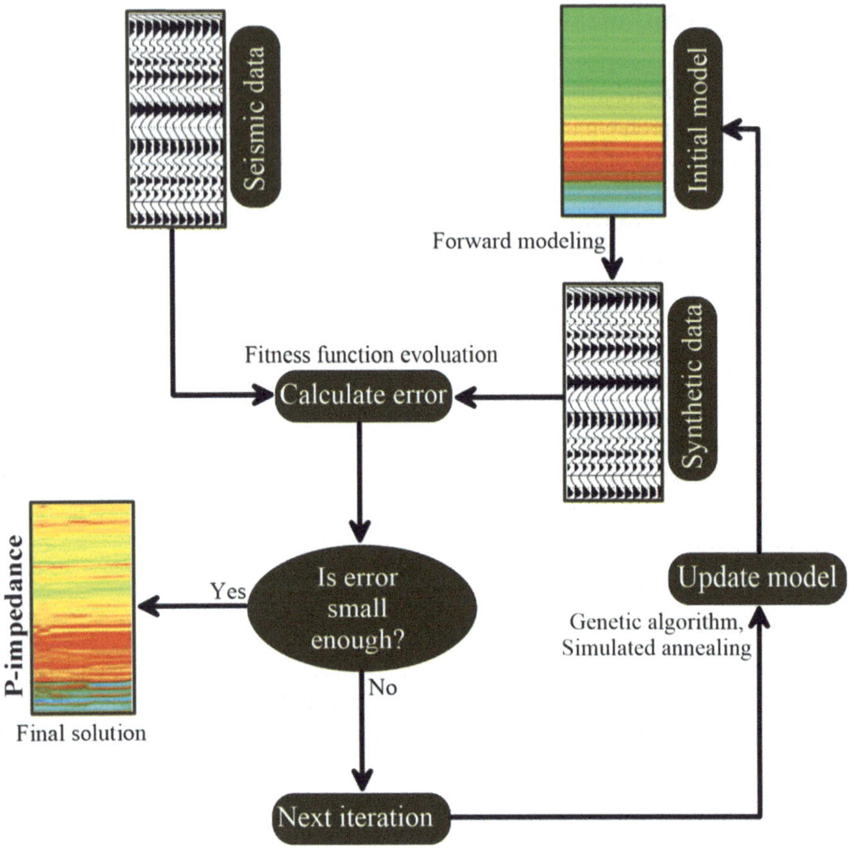

Fig. 6.3 Flow chart of global optimization methods

create randomness to the individual solution (Sivanandam and Deepa 2007; Morgan et al. 2012). These methods are explained briefly in the following subsections.

i. **Selection**

The selection is the first genetic operation carried out in the population after determining the fitness of each individual model. Based on their fitness values, the selection method is used to pair individual models. The model's fitness value is used to determine whether or not they are being chosen. Compared to the low fitness value, the models with very high value have an enormous opportunity to be chosen. This is because the fitter models are more frequently selected for reproduction. The probability of selection is directly associated with the model's fitness value (Goldberg and Holland 1988). There are six major selection types available and they are as follows.

- Stochastic uniform
- Remainder

- Roulette
- Uniform
- Tournament
- Rank.

Stochastic uniform

Stochastic uniform lays out a line in which each parent corresponds to a section of the line of length proportional to its expectation. The algorithm moves along the line in steps of equal size, one step for each parent. At each step, the algorithm allocates a parent from the section it lands on. The first step is a uniform random number less than the step size (Tran and Hiltunen 2012).

Remainder

Remainder assigns parents deterministically from the integer part of each individual's scaled value and then uses roulette selection on the remaining fractional part (Velez 2005).

Roulette

Roulette simulates a roulette wheel with the area of each segment proportional to its expectation. The algorithm then uses a random number to select one of the sections with a probability equal to its area (Yang and Honavar 1998).

Uniform

Uniform functions select parents at random from a uniform distribution using the expectations and number of parents. This results in an undirected search. Uniform selection is not a useful search strategy, but you can use it to test the genetic algorithm (Yang and Honavar 1998).

Tournament

Tournament selects each parent by choosing individuals at random, the number of which you can specify by Tournament size, and then choosing the best individual out of that set to be a parent. When Constraint parameters > Nonlinear constraint algorithm is Penalty, GA uses Tournament with size 2 (Yan-Jun et al. 2016).

Rank selection

The fitness values of the models are evaluated in the rank selection method (Baker 1987; Whitley 1989), and then they are sorted to assign a rank to each model. A model's rank can range from 0 (for the best model) to $n - 1$ (for the worst model). The selection is done in such a way that the best model in the population contributes a number of copies that are an integral number of copies that the worst model receives (the number is predetermined). Thus the choice of ranks fundamentally exaggerates the distinction between models with fitness values that are almost identical (Yuan and Zhuang 1996).

ii. **Crossover**

The crossover is the next genetic operator to be used after selecting and pairing the models. The fundamental crossover principle is the sharing of genetic information among paired models. In this manner, it can also be understood that the crossover exchanges some data between paired models and thus generates fresh models (Sen and Stoffa 2013). In geophysical problems, there are two kinds of crossover widely used namely single-point crossover and multipoint crossover. The fundamental idea of the single point crossover is that with the uniform probability distribution, the model's one-bit location is chosen at random. Then all the bits falling on the right side of the selected bit will be exchanged between two models, producing a new model. On the other hand, a bit position is randomly selected in the multipoint crossover process and all the bits right to this selected bit are exchanged among the paired models. In addition, another bit location is selected at random from the second model parameter and all other bits falling straight to this bit are swapped again. For each model parameter, this method is repeated (Goldberg and Holland 1988).

Figures 6.4 show how single- and multipoint crossovers are performed. In single-point crossover (Fig. 6.4), one bit is selected at random and all the bits between model m_i and model m_j after the crossover points are exchanged with each other. This results in the generation of two new models. On the other hand the multipoint crossover exchanges binary information between each model parameter by choosing a crossover point independently and performing crossover on a model-parameter-by-model-parameter basis (Yuan and Zhuang 1996).

Once a crossover site has been randomly selected, it is decided on the basis of the probability of crossover whether or not crossover will be performed; i.e. the crossover rate is controlled by a probability px specified by the GA designer. A high probability of crossover implies that crossover between the paired models or between the present model parameter of the paired models is likely to happen in the event of multipoint crossover (Boschetti et al. 1996; Sen and Stoffa 2013; Maurya et al. 2017).

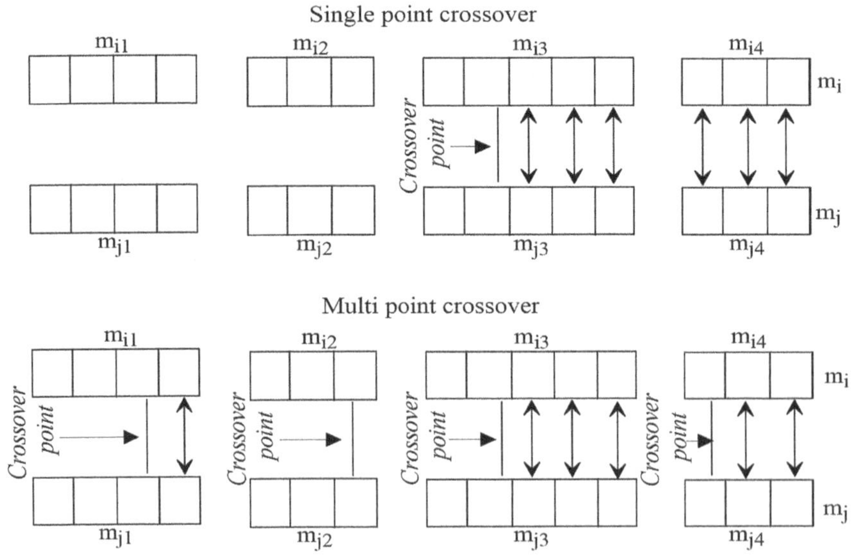

Fig. 6.4 Single and multi-point crossover process

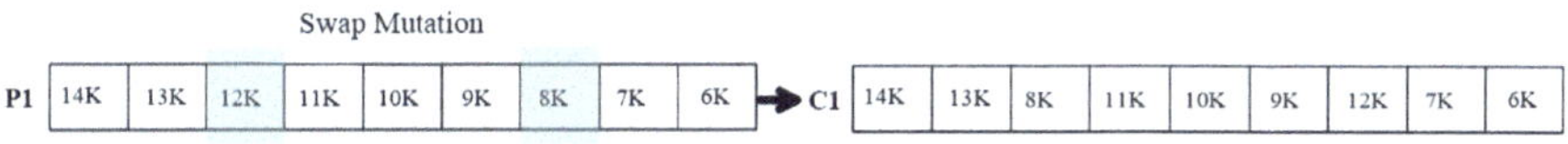

Fig. 6.5 Basic concept of mutation where K represents 1000

iii. **Mutation**

The last genetic operator is mutation. The mutation is a process used to create randomness to the crossover. Generally, the process of mutation is performed during the crossover process. The mutation rate which is used to decide the number of walks in the model space is specified by a probability that is determined by the algorithm designer. The low mutation probability means fewer walks in the model space and the conversion of the problem will be very fast (Holland 1975; Deb et al. 2002). On the other hand, the high mutation probability will result in a large number of random walks in the space but this can take more time to converge the algorithm (Mallick 1995; Aleardi et al. 2016). Figure 6.5 shows the basic concept of the mutation. One new solution (children) is selected from the crossover and treated as parents to perform mutation. For this purpose, two mutation points are selected and interchanges with each other to produce randomness to the solution (Mallick 1995).

The inversion based on genetic algorithm does not give an absolute subsurface impedance model, it gives a relative variation of acoustic impedance. The practical process of the genetic algorithm is as follows.

Mathematical example

The process can understand in a specific way with the examples. Let us consider a function

$$f(x) = \frac{\cos(3\pi x)}{x}, \text{ where} x \in [0.11.1] \tag{6.21}$$

This function graphically can be presented as follows (Fig. 6.6).

The function has a local and global maximum and local and global minimums and they are highlighted. The function has a global minimum fall at x = 0.30 and the corresponding function value is −3.1705. To demonstrate the capability of a genetic algorithm to find the solution to the problem, the method is applied to Eq. 6.21. The estimated results are as follows.

$$x = 0.296878161902085$$

And

$$f(x) = -3.17151688802542$$

This is very close to the real solution and describes the high performance of the algorithm.

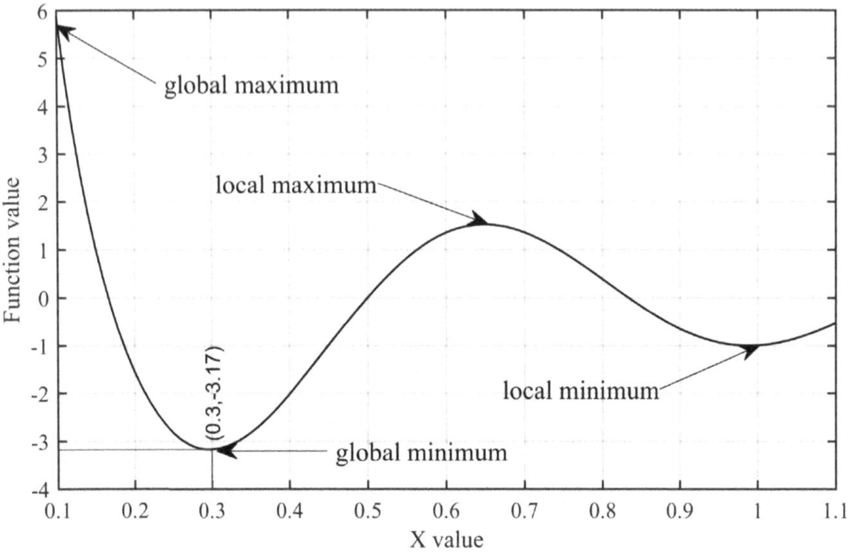

Fig. 6.6 Plot of function presented in Eq. 6.21

Synthetic example
The above-discussed example is presented in a mathematical case although the GA largely used in seismic inversion purpose. Hence, an example is present to demonstrate the use of GA for seismic inversion. In this regard, the parameters are optimized by applying the first synthetic data and then real composite trace. A seven-layer earth model is assumed with acoustic impedance 8050, 10,000, 7260, 9120 and 11,440, 7820, 10,250 m/s $*$ g/cc. The result for a seven-layer Earth's model is illustrated in Fig. 6.7. The first track of Fig. 6.7 shows assumed subsurface seven-layer earth model, second track depicts corresponding synthetic seismogram generated by the convolution is 40 Hz Ricker wavelet with earth reflectivity, the same trace shown 8 times to looks like data and third track depicts comparison of modeled (dot line) impedance with observed impedance (solid line). From the figure, it can be examined that the inverted acoustic impedance is very close to the modeled acoustic impedance and depicts the good performance of the algorithm. To quality check of inverted results, crossplot between inverted and original data is generated and shown in Fig. 6.8. The red solid line in the figure represents best-fit line. It can be noticed that the scatter points are very close to the best fit line and hence indicates that the inverted results are in the vicinity of real value and indicates the good performance of the algorithm. The correlation is estimated to be 0.96 between modeled and observed impedance which is very high and supports the high performance of the algorithm.

Figure 6.9 shows visualization of error minimization. One can notice that the error decreases exponentially with the iteration and reached to minimum value of 0.0069 after 100 iterations. The termination criterion is set to be 100 iterations by considering

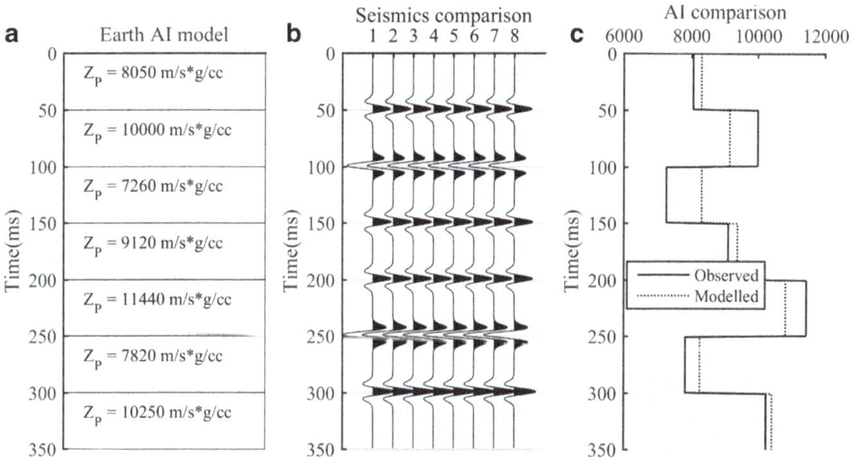

Fig. 6.7 Demonstrate seismic inversion results for synthetic data case whereas **a** represents a geological model, **b** depicts synthetic trace generated over geological model and **c** compares inversion results with the original value

Fig. 6.8 Crossplot of modeled and actual impedance for synthetic data inversion

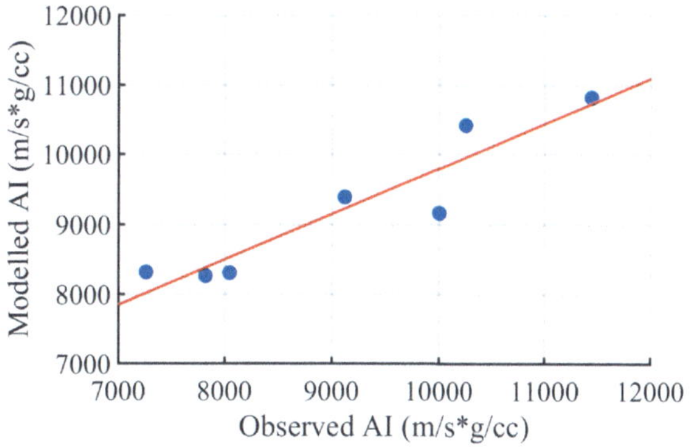

synthetic data case. The synthetic example contains 7 layers only and hence one can compare layer by layer the inversion results. This comparison is not easy for real data cases as the subsurface contains a large number of layers. Table 6.1 shows a quantitative comparison of inverted results with original value for each layer. From the table, one can notice that the modeled impedance is quite close to the observed impedance. The change percentage between observed and modeled data points is also estimated and found very low (<5.967% average).

Real data example

Now, we are in a position to move towards real data example as the mathematical and synthetic example shows satisfactory results. These analysis always important as the global optimization methods implemented by many parameters and hence they become very crucial to take input. One other reason is that this global optimization

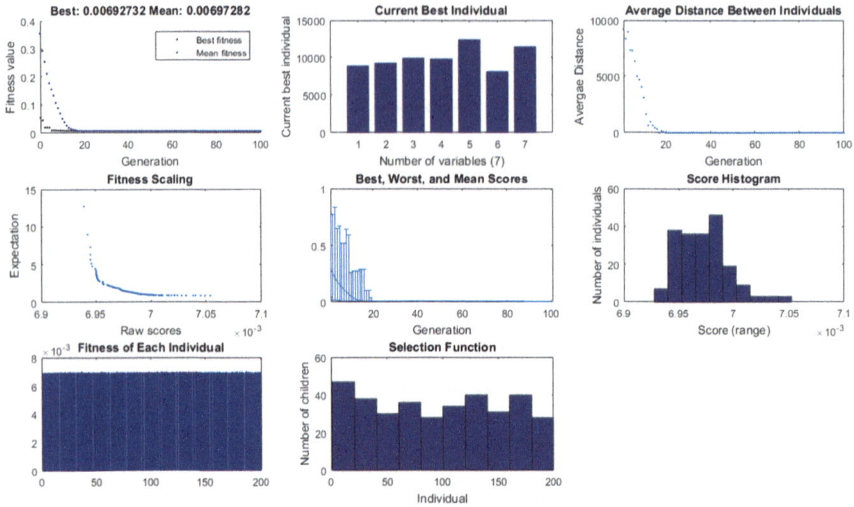

Fig. 6.9 Visualization of errors for the inversion of synthetic data with genetic algorithm

Table 6.1 Comparison of observed and modeled AI for synthetic data case

S. No.	Layers	Observed AI	Modeled AI	Change (%)
1.	Layer 1	8050	8300.93	3.12
2.	Layer 2	10,000	9162.50	8.38
3.	Layer 3	7260	8310.48	14.47
4.	Layer 4	9120	9393.93	3.00
5.	Layer 5	11,440	10,817.03	5.45
6.	Layer 6	7820	8261.50	5.65
7.	Layer 7	10,250	10,423.94	1.70

is much time taking and hence one needs to optimize parameters for small datasets before moving for large data sets. The real data is taken from the Blackfoot field, Alberta, Canada. The GA is applied to estimate acoustic impedance in the inter-well region. The real data example is divided into two parts, first, one composite trace near to well 08-08 is extracted and GA is performed and in the second part, the entire seismic section is transformed into the AI using GA technique.

For real data case, a real coded GA is implemented where an individual is represented by an array of real numbers and these are velocity, density, and thickness for each layer. In the first step, define a search width for each of the model parameters, which is obtained from a priori information. The lower bound is decided to be used 2250 m/s * g/cc and the upper bound is 21,000 m/s * g/cc for impedance which is a reasonable range for seismic reflection data. GA starts with the generation of a random population of 200 chromosomes. Using forward modeling procedure the synthetic seismograms are generated for each individual chromosomes and corresponding fitness value is evaluated. On the basis of these fitness values, two best solutions/chromosomes have been selected to perform the genetic operation.

Figure 6.10 shows comparison of inverted impedance and real impedance from well log. Track 1 shows extracted seismic trace, the same trace plotted 8 times to look like data, track 2 shows extracted wavelet from well log data which is used for forwarding modeling and track 3 shows comparison of model, original and inverted impedances. From track 3 (Fig. 6.10), one can notice that the inverted curves are very close to the original curves. The correlation between them is noticed to be 0.88 which is quite good by considering noise in the data.

Again, for the quality check of the inversion results, a cross plot between inverted and original impedances is generated and displays in Fig. 6.11. A best-fit line is fitted to the data. Now, one can observe that the distribution of scatter points lies close to the best fit line which indicates that the inverted results are close to the original one. The statistical analyses of inverted results are presented in Table 6.2. The table shows the minimum and maximum AI estimated for composite trace. The variation of error with the iteration is shown in Fig. 6.12. The stopping criterion is set to be 900 iterations. One can increase the iteration number to get more close to the real solution but here we have decided to use 900 iterations as a stopping criterion to

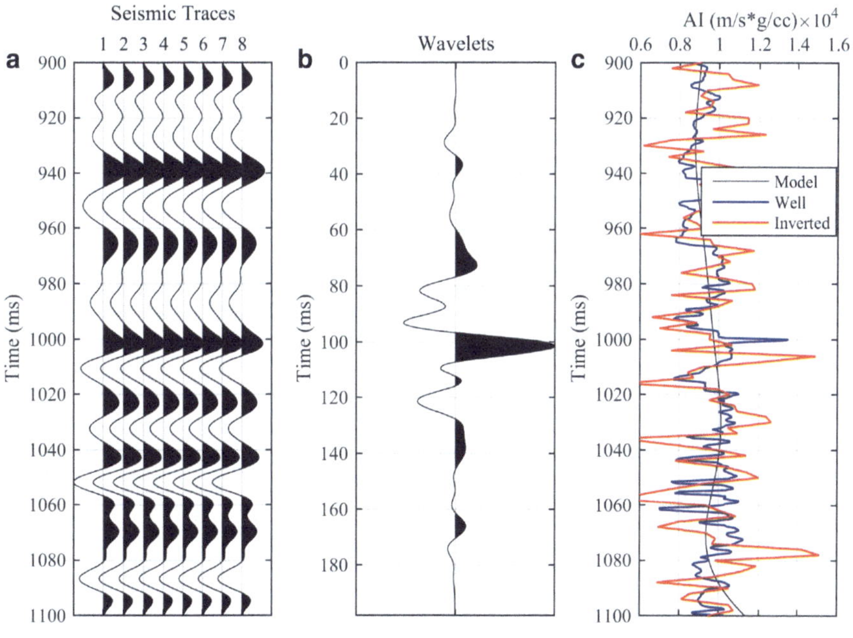

Fig. 6.10 Inversion results for the composite trace from the Blackfoot field, where **a** represents seismic trace (one trace plotted 8 times), **b** shows extracted wavelet and **c** compares inverted impedance, model impedance, and original impedance

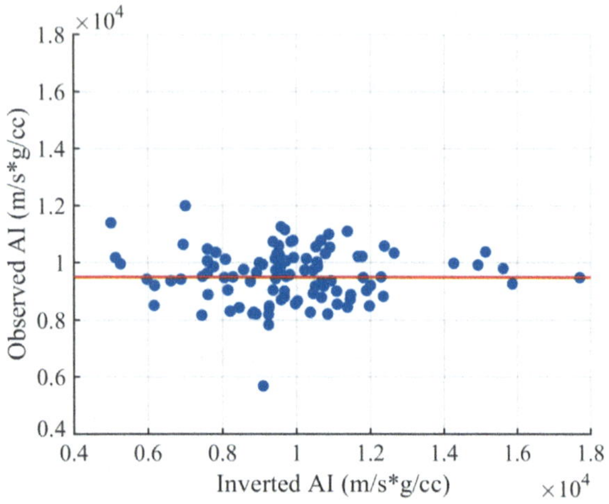

Fig. 6.11 Crossplot of original and inverted impedances for composite trace

Table 6.2 Statistical information for composite trace

S. No.	Parameters	Observed	Inverted	Change (%)
1.	Minimum AI	7035	5001	28.91
2.	Maximum AI	13,510	17,110	21.04
3.	Mean AI	9567	9840	2.77

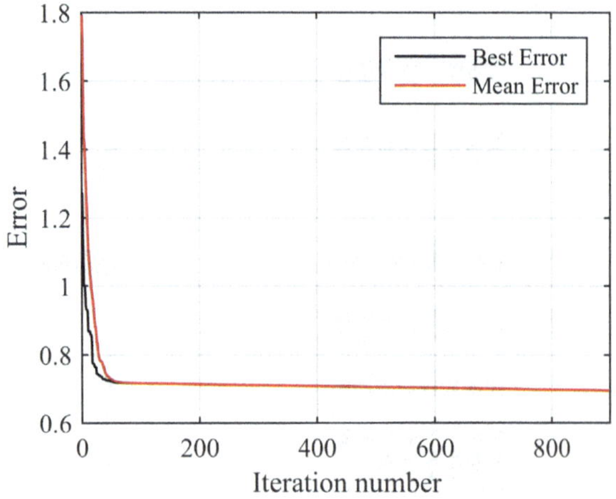

Fig. 6.12 Variation of best and mean error with the iteration number

optimize time and memory. From the error analysis, one can notice that it decreases exponentially with iteration.

The analysis of composite trace shows satisfactory results and therefore, the GA algorithm is applied to the entire seismic section to estimate AI in the inter-well region. Figure 6.13 (top) shows a cross-section of seismic section which is used

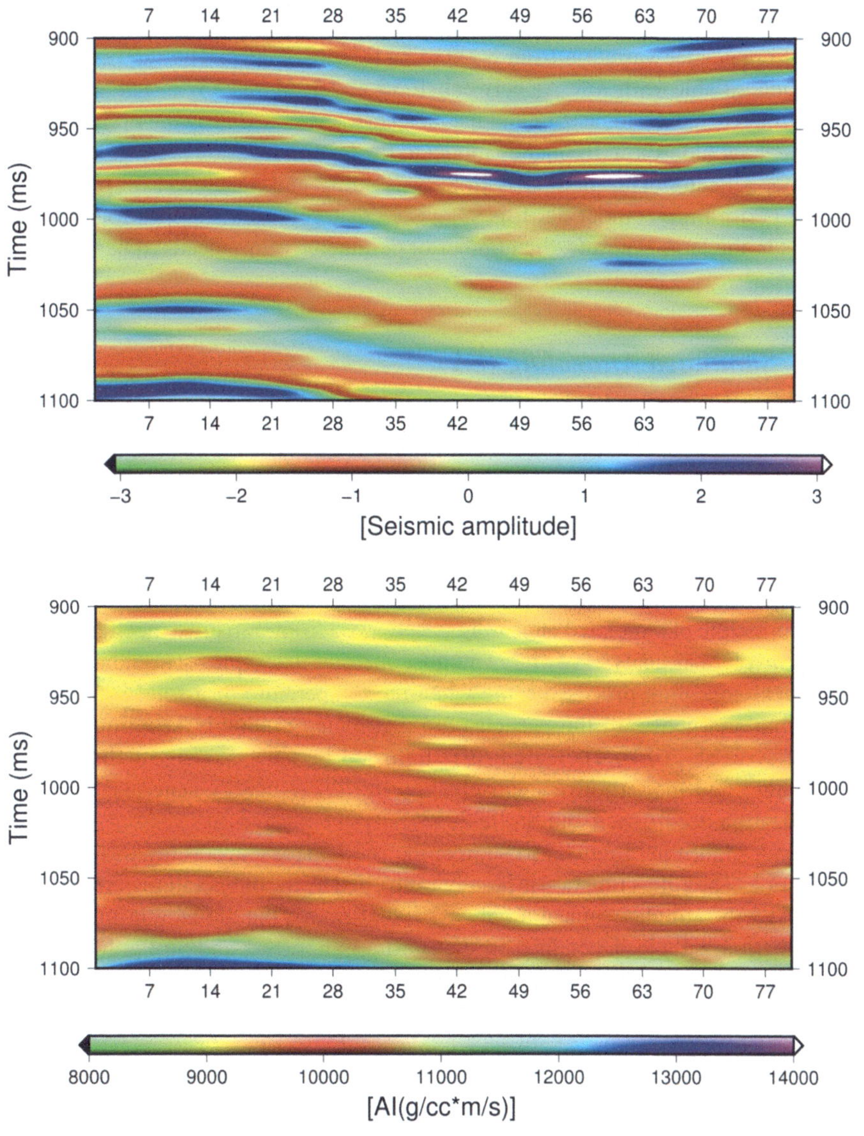

Fig. 6.13 Cross-section of seismic data (top) from the Blackfoot field and inverted impedance (bottom) derived using GA

as input for GA and Fig. 6.13 (bottom) shows cross-section of inverted acoustic impedance (*AI*) at inline 27. The inverted section shows higher resolution as compared with the seismic data and shows a clear boundary of subsurface layers. From the inverted AI section, the thinner layer cannot be interpreted because of seismic bandwidth and also because the algorithm does not include constraints from well log information as other inversion methods used. The section shows that the area have acoustic impedance variation from 8000 to 14,000 m/s * g/cc. The sand channel is clearly identifiable with low AI between 1055 and 1065 ms TWT. The minimum RMS error is achieved to be 0.12 and the mean error as 0.20 during the optimization.

One can include a low-frequency model to the inverted section to get information on deeper reflector. The method is time-consuming and hence can be performed on high performing computing. It is suggested that before the application of these methods to the large datasets, the number of parameter values needs to be tested.

6.4.2 Simulated Annealing (SA)

The Simulated Annealing (SA) is another type of global optimization method used to estimate the global optimum solution of a function using temperature cooling principle. The methods are very useful to the geophysical problem and some nice examples are presented in Kirkpatrick et al. (1983), Geman and Geman (1987), Davis (1987), Van Laarhoven and Aarts (1987), Aarts and Korst (1988), Sen and Stoffa (1991) and Afanasiev et al. (2014).

The principle of simulated annealing is borrowed from statistical mechanics. The method involves an analysis of a large number of atoms in samples of liquids or solids. A physical annealing process occurs when the temperature of a solid is increased by heating and hence all the particles are distributed randomly in a liquid phase (Afanasiev et al. 2014). Thereafter, it is followed by slow cooling of the particle temperature and they arrange themselves in the lower energy state and resultant crystallization occurs (Sen and Stoffa 2013).

The SA algorithm attempts to sample the model space according to the Gibbs distribution. The Gibbs function is largely controlled by the temperature of the model. That's why the simulate annealing always look towards the global minimum of an error function without depending on the starting point (Sen and Stoffa 2013). The Gibbs distribution can be written as follows.

$$P(E_i) = \frac{\exp\left(-\frac{E_i}{KT}\right)}{\sum_{j\in S}\exp\left(-\frac{E_j}{KT}\right)} = \frac{1}{Z(T)}\exp\left(-\frac{E_i}{KT}\right) \tag{6.22}$$

where the model state is i, the energy of the medium is represented as E_i, the set S consists of all possible configurations, K is Boltzmann's constant, T is the temperature, and $Z(T)$ is the partition function/energy. This partition energy can be written as follows.

$$Z(T) = \sum_{j\in S}\exp\left(\frac{E_j}{KT}\right) \tag{6.23}$$

The simulated annealing algorithm starts with an initial temperature that is termed as initial model m_0, and the energy associated with this model can be expressed as $f(m_0)$. The SA generate new model, let say it is m_n and the energy associated with this new model can be expressed as $f(m_n)$. Thereafter, compare the new energy $f(m_n)$ with the earlier energy state i.e. $f(m_0)$. If the new energy state is lower than that of the starting model, then the new model is accepted without any condition. On the other hand, if the new energy state is larger than that of the initial model, then the new model is accepted with the following probability (Ma 2002).

$$p = \exp\left[-\frac{(f(m_n) - f(m_0))}{T}\right] \tag{6.24}$$

In Eq. 6.24, T is a control parameter called the acceptance temperature (Ma 2002). As described above, the same process is repeated a large number of times, with the annealing temperature gradually decreasing according to a cooling schedule. A stopping criterion is set to stop the algorithm (Sen and Stoffa 1991; Goffe et al. 1994; Ma 2002).

There are many simulated annealing algorithms available and some of them are as follows.

- Metropolis algorithm
- Heat bath algorithm
- Fast simulated annealing.

Metropolis algorithm

Metropolis algorithm of simulated annealing is developed by Metropolis in 1953 (Metropolis et al. 1953). Since then, this Metropolis algorithm with annealing is used in a verity of problems. The basic principle of the metropolis algorithm can understand as following.

Let us assume a starting model m_i with energy $E(m_i)$. Further, a small perturbation to the initial model gives a new model which is represented by m_j with energy $E(m_j)$. Mathematical relationship between the initial model and the new model can be written as follows.

$$m_j = m_i + \Delta m_i \tag{6.25}$$

Now, lets the difference in energy between the initial model and newly generated model is presented by ΔE_{ij}. This energy difference can be given by the following equation.

$$\Delta E_{ij} = E(m_j) - (m_i) \tag{6.26}$$

The new model is better or not from the previous one is decided based on the value of ΔE_{ij}. If the energy difference is zero ($\Delta E_{ij} = 0$), then the new model is always get accepted. On the other hand, if the energy difference is greater than zero ($\Delta E_{ij} > 0$), then the new model gets accepted with the following probability.

$$P = \exp\left(-\frac{\Delta E_{ij}}{T}\right) \tag{6.27}$$

The previous principle of acceptance is known as the criterion of Metropolis. If the generation-acceptance process is repeated many times at each temperature, it can be shown that at each temperature the thermal equilibrium can be achieved. If the temperature is progressively reduced following a cooling scheme such that at each temperature thermal equilibrium has been reached, then the global minimum energy state can be reached in the limit as the temperature approaches zero (Geman and Geman 1984).

We begin with a reference model in a local search technique, and a fresh model is adopted if and only if $\Delta E_{ijI} = 0$; i.e. it always searches downhill direction. In SA, as described by the Metropolis algorithm, each model has a finite likelihood of acceptance despite the fact that $\Delta E_{ij} > 0$ is possible. Local optimization can thus be trapped in a local minimum that may be in the vicinity of the starting model, whereas SA has a finite likelihood of jumping out of the local minima (Sen et al. 1995; Ma, 2002). However, as the temperature approaches zero, only moves showing improvement over the previous test are likely to be accepted, and the algorithm decreases to a greedy algorithm in the limit $T \to 0$.

Heat bath algorithm

As described in the previous section, the Metropolis algorithm can be considered a two-step procedure in which a random move is made first, and it is then decided whether or not to accept the move. So many of these movements are going to be dismissed. The rejection-to-acceptance ratio is very large, particularly at low temperatures. This scenario has been remedied by several algorithms—one of which is the thermal bath algorithm (Rebbi 1987; Geman and Geman 1984). Unlike the Metropolis algorithm, the heat bath algorithm is a one-step operation that tries to prevent a high rejection-to-acceptance ratio by calculating the relative acceptance probabilities of each move before any random assumptions are made. The method simply produces weighted selections that are always accepted. The previous declaration may be somewhat misleading as a weighted guess is made only at significant cost, as is shown by a detailed analysis of the algorithm.

Consider a model vector m made up of parameters of the N model. Further, assume that each m_i can attain maximum M values. This can be achieved by assigning smaller boundaries (m_i^{min}) and higher boundaries (m_i^{max}) and performing a search increment of m_i for each parameter of the model such that

$$M = \frac{m_i^{max} - m_i^{min}}{\Delta m_i} \tag{6.28}$$

From Eq. 6.28, it can be noted that the parameter M can take different values for different model parameters. However, we will suppose that M is the same for all model parameters without loss of generality. This results in a model space made up of MN models. From this point in time, it is convenient to present the model parameter by m_{ij} where $i = 1, 2, \ldots, N$ and $j = 1, 2, \ldots, M$.

The algorithm starts with an initial guess model m_0 and then each parameter of the model is sequentially visited. The following marginal probability density function (pdf) is assessed for each parameter of the model.

$$\widehat{P}\left(\bar{m}|m_i = m_{ij}\right) = \frac{\exp\left[-\frac{E\left(\bar{m}|m_i=m_{ij}\right)}{T}\right]}{\sum_{k=1}^{M}\left[-\frac{E\left(\bar{m}|m_i=m_{ik}\right)}{T}\right]} \tag{6.29}$$

The sum is over M values permitted by the model parameter and $E\left(\bar{m}|m_i = m_{ij}\right)$ is the energy of the model vector $\bar{m}$ whose model parameter has a value m_{ij}. Then a value from the previous allocation is taken. Essentially, a random number is taken from the uniform distribution $U[0, 1]$ is mapped to the pdf given in Eq. 6.29. To perform this, first, we need to compute the cumulative probability C_{ij} from $\widehat{P}\left(\bar{m}|m_i = m_{ij}\right)$. This can write as follows.

$$C_{ij} = \sum_{k=1}^{j} \widehat{P}(\bar{m}|m_i = m_{ik}),\ J = 1, 2, \ldots, M. \tag{6.30}$$

Thereafter, one can draw a random number r from the uniform distribution. At $j = k$, where $C_{ij} = r$, we select $m_{ij} = m_{ik} = m_i$. The distribution is almost uniform at high temperatures and therefore every parameter of the model is almost equally probable to be selected. Only the peak corresponding to the model parameter that most dominates the error function at very low temperatures.

The new value of the model parameter thus selected and replaces the old value of the corresponding model parameter in the model vector, and for each model parameter, the process is repeated sequentially. After having examined each parameter of the model once, we may have a model different from the starting model. This is one iteration of the algorithm consisting of assessing the error function $N \times M$ times. Unlike Metropolis SA, the heat bath does not involve testing for accepting a model after it has been produced, but this algorithm involves significant computational cost. When the problems have a large number of variables, it is faster than the Metropolis SA (Vestergaard and Mosegaard 1991).

Fast simulated annealing (FSA)

Simulated annealing is a stochastic system for looking through the ground state. Fast simulated annealing (FSA) is a semi-neighborhood search and comprises of intermittent long bounces. SA's Markov chain model can be used to prove that the asymptotic convergence given by the Boltzmann distribution to a stationary distribution can be achieved, and the global minimum energy state has a probability of 1 when the temperature goes to zero. However, the speed at which the temperature is reduced depends on whether or not the global minimum energy state is reached. It means it relies on the cooling schedule. Geman and Geman (1984) demonstrated that the following cooling plan provides a required and adequate condition for convergence to the worldwide minimum for SA.

$$T(k) = \frac{T_0}{\ln(k)} \tag{6.31}$$

In Eq. 6.31, $T(k)$ is the temperature at iteration k and T_0 is a sufficiently high starting temperature. The above discussed cooling schedule is much time taking and hence they need much time to converge at the optimum solution.

To overcome the problem, Szu and Harley (1987) proposed a new technique that is called fast simulated annealing (FSA). This algorithm is very similar to the Metropolis algorithm discussed above. Has only difference is that it uses Cauchy distribution of the model parameters rather than flat distribution. If a Cauchy distribution attained high temperature then the distribution allows far selection from the present position whereas if it attained low temperature then the close neighborhood selection occurs. The advantage of using Cauchy distribution over Gaussian distribution is that it has a flatter tail and therefore has a better chance of getting out of local minima. Szu and Hartley (1987) also showed that the convergence cooling plan is no longer logarithmic and the temperature plan is now inversely proportional to the iteration number.

$$T(k) = \frac{T_0}{k} \tag{6.32}$$

These are the basic principle of a different kind of simulated annealing algorithm but the get the real picture, some of them are applied to the synthetic as well as real data sets. In the above-discussed annealing function, the fast annealing is utilized in these examples.

Synthetic data example

To minimize error for synthetic as well as real data, the following merit function is utilized.

$$Error(E) = \frac{1}{n}\sqrt{\sum_{i=1}^{n}\left(S_{obs}^{i} - S_{mod}^{i}\right)^2} + \frac{1}{n}\sqrt{\sum_{i=1}^{n}\left(Z_{obs}^{i} - Z_{mod}^{i}\right)^2} \tag{6.33}$$

The algorithm is utilized to estimate acoustic impedances. The synthetic seismograms are produced by considering 7 layer model with acoustic impedances of 8050, 10,000, 7260, 9120 and 11,440, 7820, 10,250 m/s $*$ g/cc. It is presumed that seismic information can be approximated in laterally homogeneous acoustic media by converting reflective coefficients with a known source wavelet. The convolution model assumes plane-wave propagation across the limits of horizontally homogeneous layers and does not take into consideration the impacts of geometric divergence, elastic absorption, wavelet dispersion, loss of transmission, conversion of mode, and numerous reflections. The seismic information must be processed to eliminate these impacts and restore plane-wave amplitudes of main P-wave reflections in order for a convolution model to be valid. After generating synthetic datasets, now, one can perform Simulate Annealing to minimize the error between observed and modeled data.

The search width is restricted as the lower limits (LB) are (7500, 9000, 6500, 8500, 11,000, 7000, 9500 m/s * g/cc) and the upper limits (UB) are (9500, 10,500, 7500, 1000, 12,000, 8000, 11,000 m/s * g/cc). The advantage of using LB and UB is creating a window within that the search for optimum solution performed within the interval to reduce time and space. The Simulated Annealing algorithm is used to restore acoustic impedance in the synthetic datasets. The seismic inversion method starts by using the original model parameters to calculate the objective function. From either a random model or a macro model a starting model is traced. According to Metropolis criteria, a fresh model is adopted or dismissed. Large numbers of iterations are carried out at a specified temperature. Then the temperature will be reduced until eventually the convergence criterion is met. The outputs at each sample interval are the optimized acoustic impedance values.

Figure 6.14 shows inversion results for synthetic data. The first track of Fig. 6.14 displays the geological subsurface model used to generate synthetic data, the second track demonstrates the respective synthetic traces, the same trace plotted 8 times to look like data and the third track shows Simulated annealing reversal outcomes. The inverted trace is very close to the original curves. The correlation between observed and modeled impedance is noted to be 0.99, which is very high and depicts the good performance of the algorithm. The estimated P-impedances can be seen to match their real models quite nicely. From Fig. 6.14, track 3, it is evident that the original impedance can get back quite nicely using a Simulated annealing technique.

Layer by layer comparison of inverted impedance is presented in Table 6.3. The table demonstrates comparison of observed and modeled acoustic impedance estimated by the simulated algorithm. From this comparison, it can be seen that the modeled AI is very close to the observed AI for all 7 layers and hence the change

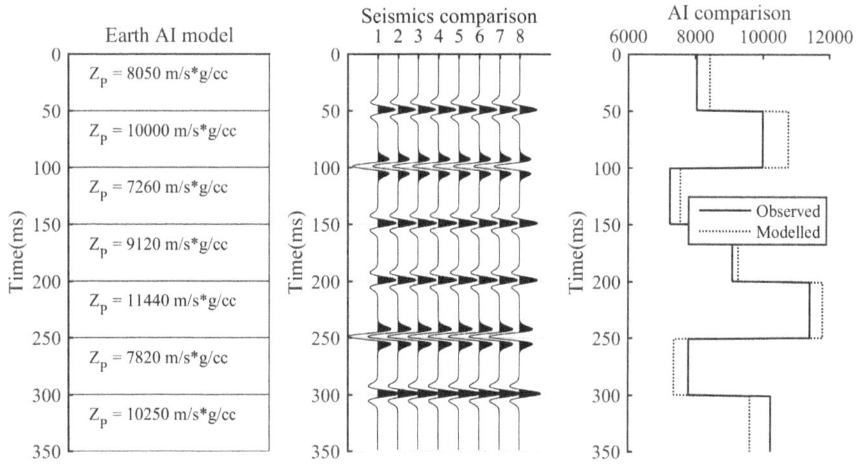

Fig. 6.14 Demonstrate seismic inversion results for synthetic data case whereas track 1 represents the geological model, track 2 depicts synthetic trace generated over geological model and track 3 compares inversion results with the original value

Table 6.3 Comparison of observed and modeled AI for synthetic data case

S. No.	Layers	Observed AI	Modeled AI	Change (%)
1.	Layer 1	8050	8436.99	4.81
2.	Layer 2	10,000	10,767.11	7.67
3.	Layer 3	7260	7569.52	4.26
4.	Layer 4	9120	9290.67	1.87
5.	Layer 5	11,440	11,819.47	3.32
6.	Layer 6	7820	7382.39	5.6
7.	Layer 7	10,250	9636.29	5.99

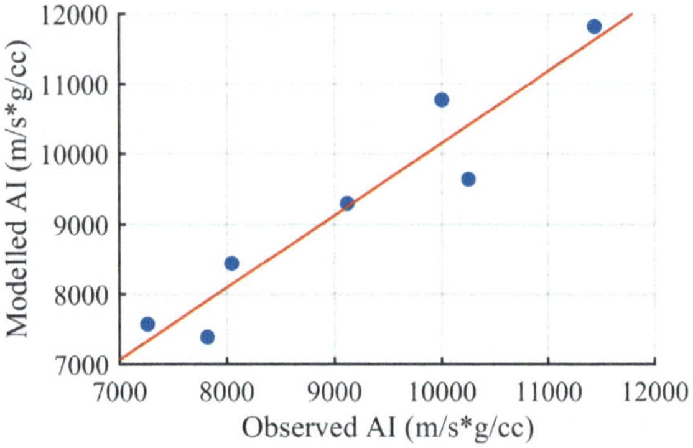

Fig. 6.15 Cross plot of original and modeled impedance for synthetic data case

percentage is very low (<4.788, average). Figure 6.15 shows a cross-plot of modeled and observed acoustic impedance. This cross plot shows visually how close both results are. A best-fit line is fitted to the scatter data. From Fig. 6.15, it is visible that the scatter points are very close to the best fit line and show excellent SA algorithm efficiency. Figure 6.16 demonstrates annealing error assessment. The plots show that the error decreases with iteration but exponentially as in the case of a genetic algorithm. The stopping criterion is set to be 500 iterations. The simulated annealing is comparatively faster than the genetic algorithm.

A comparison of results found in the genetic algorithm and simulated annealing for synthetic data case has been performed. Figure 6.17 shows a cross plot of observed and modeled acoustic impedance for both the case. It's noticeable from Fig. 6.17 that the scatter points are very near to the best fit line for both instances (GA and SA) and show excellent algorithm efficiency. The close examination of the cross plot indicates that the data points from simulated annealing are very close to the best fit line whereas, for the genetic algorithm, the data points are scattered relatively. The correlation coefficient for SA and GA is anticipated to be 0.987 and 0.937 respectively. This analysis indicates that the inversion results for synthetic data cases are slightly better for simulated annealing as compared with the genetic algorithm. This is not always the case, it largely depends on data parameters and input data.

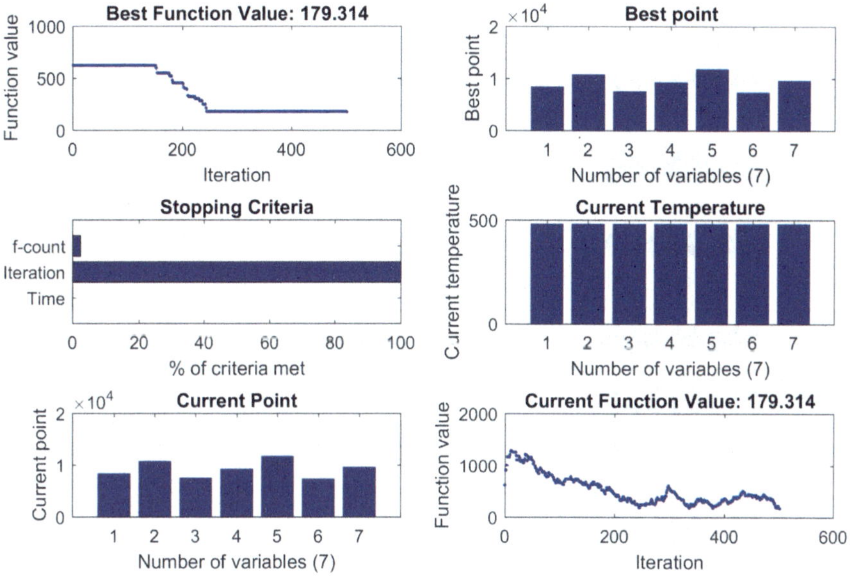

Fig. 6.16 Variation of errors with iteration for synthetic data

Fig. 6.17 Crossplot of original and inverted acoustic impedance for synthetic data. The figure includes both results estimated from simulated annealing and genetic algorithm

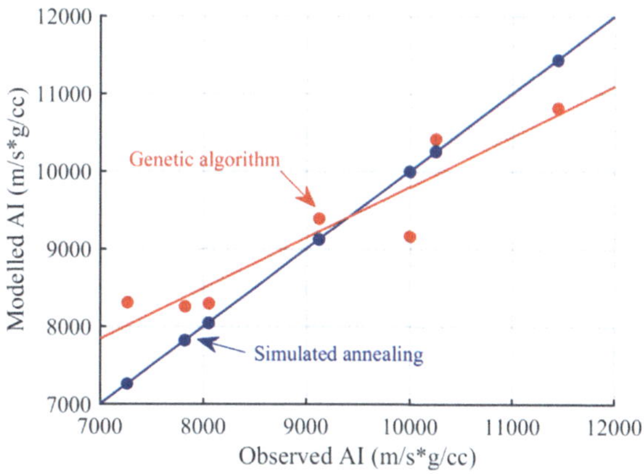

Real data example

The other example presented here is from the real data from the Blackfoot field, Alberta, Canada. This example is presented to estimate acoustic impedance in the inter-well region using seismic reflection data as input. The application to the simulated annealing is conducted in two phases. First, extracts a composite seismic trace near the well location and SA is conducted on it. Figure 6.18 indicates the outcome of the assessment of the composite trace. Track 1 of Fig. 6.18 indicates true seismic trace close to well 08-08 (the same trace plotted 8 times), track 2 represents

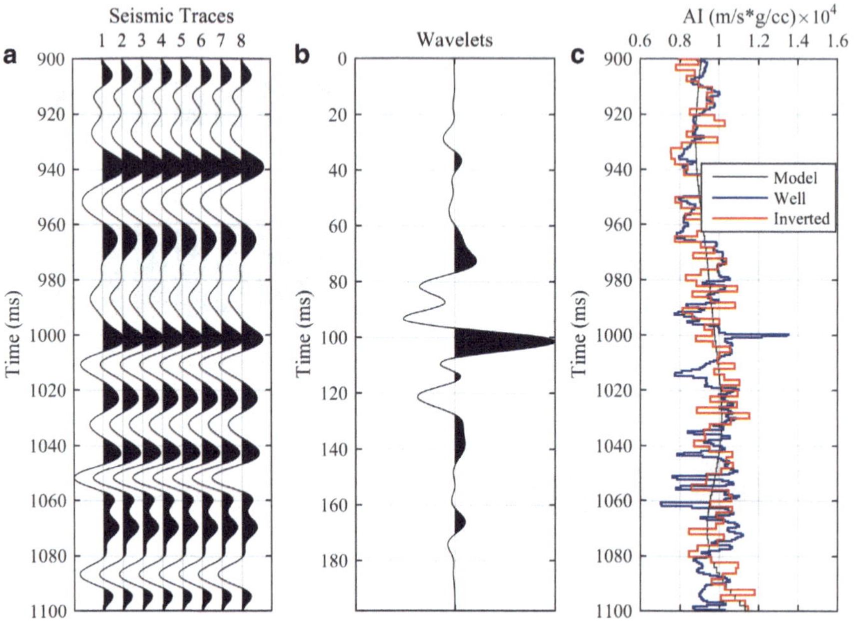

Fig. 6.18 Inversion results for the composite trace from the Blackfoot field, where **a** represents seismic trace (one trace plotted 8 times), **b** shows extracted wavelet and **c** compares inverted impedance, model impedance, and original impedance

extracted wavelet used to generated modeled trace and track 3 demonstrates an estimated inversion outcome using simulated annealing algorithm. By examining the inverted outcomes, one can see that the inverted curve is quite nicely following the original impedance from well log. The correlation between estimated and observed AI is estimated to be 0.95 which is quite high and indicates good inversion results.

A cross plot between inverted impedance and the original impedance is produced and displayed in Fig. 6.19 for quality check of inversion results. In addition, the best fit line is fitted to the data. From the cross plot, one may notice that the scatter points fall close to the best fit line and show excellent algorithm efficiency. The statistical analysis of composite trace is given in Table 6.4.

Once acceptable outcomes have been obtained from the composite trace, the SA technique is applied to the entire seismic segment to assess the acoustic impedance of the subsurface in the inter-well region. Figure 6.20 (top) demonstrates seismic input information (Inline 65) from the field of Blackfoot, Alberta, Canada and Fig. 6.20 (bottom) shows inverted results at inline 65. The inverted section demonstrates comparatively greater resolution and offers the layer property while seismic information only offers the property of the interface. The impedance varies from 7500 to 13,500 m/s * g/cc. The inverted acoustic impedance section highlights sand channel between 1055 and 1065 ms TWT.

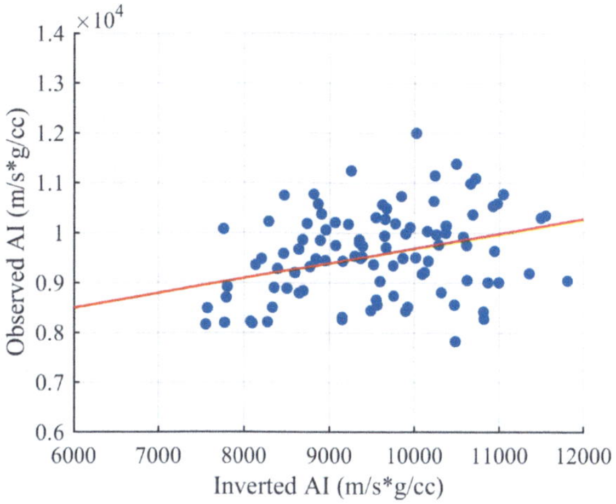

Fig. 6.19 Crossplot of inverted and actual impedance for composite trace. The inverted impedance is estimated using SA

Table 6.4 Statistical information for composite run

S. No.	Parameters	Observed	Inverted	Change (%)
1.	Minimum	7035	7551	7.33
2.	Maximum	13,510	11,810	12.58
3.	Mean	9567	9537	0.31

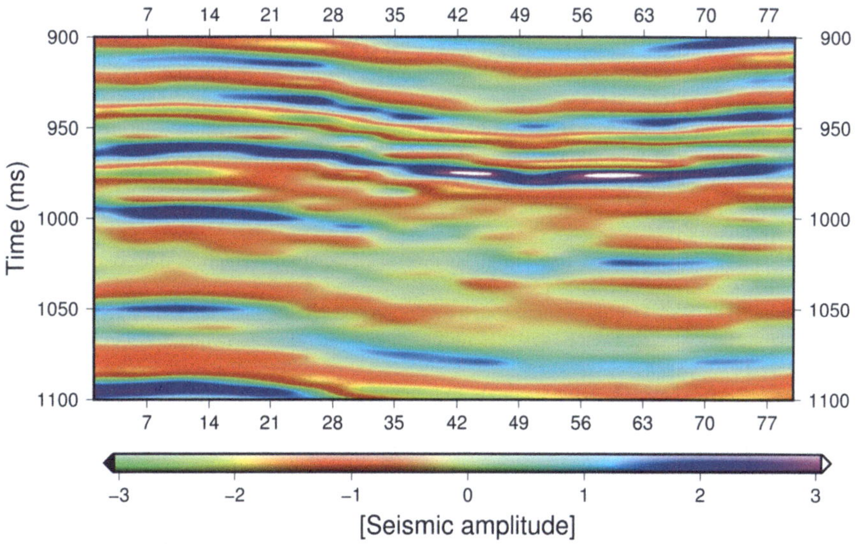

Fig. 6.20 Cross-section of seismic data (top) from the Blackfoot field and inverted impedance (bottom) derived using SA

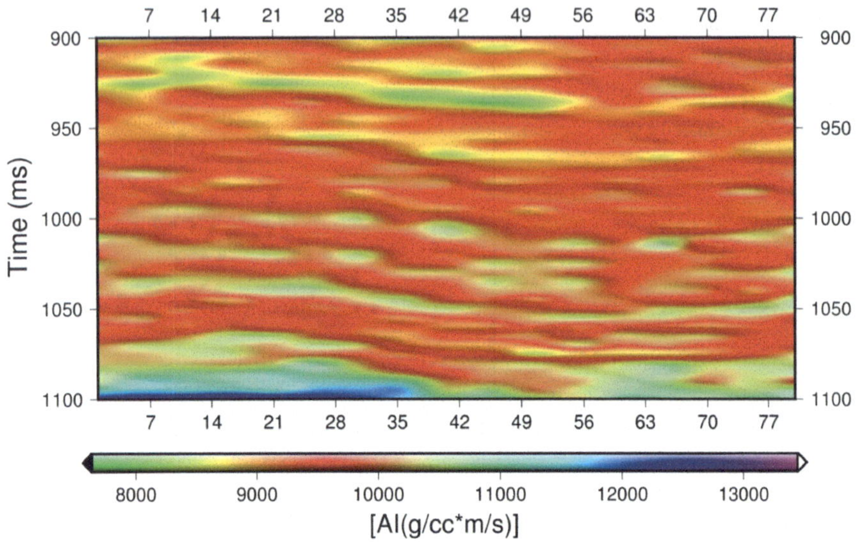

Fig. 6.20 (continued)

References

Aarts E, Korst J (1988) Simulated annealing and Boltzmann machines

Afanasiev MV, Pratt RG, Kamei R, McDowell G (2014) Waveform-based simulated annealing of cross-hole transmission data: a semi-global method for estimating seismic anisotropy. Geophys J Int 199(3):1586–1607

Aleardi M, Tognarelli A, Mazzotti A (2016) Characterization of shallow marine sediments using high-resolution velocity analysis and genetic-algorithm-driven 1D elastic full-waveform inversion. Near Surf Geophys 14(5):449–460

Baker JE (1987) Reducing bias and inefficiency in the selection algorithm. Proc Second Int Conf Genet Algorit 206:14–21

Boschetti F, Dentith MC, List RD (1996) Inversion of seismic refraction data using genetic algorithms. Geophysics 61(6):1715–1727

Bremermann HJ (1962) Optimization through evolution and recombination. Self-org Syst 93:106

Bremermann H, Oehme R, Taylor J (1958) Proof of dispersion relations in quantized field theories. Phys Rev 109(6):2178

Broyden CG (1967) Quasi-Newton methods and their application to function minimization. Math Comput 21(99):368–381

Davis L (1987) Genetic algorithms and simulated annealing. Morgan Kaufmann Publishers Inc., San Francisco, CA, United States

Deb K, Pratap A, Sameer A, Meyarivan TAMT (2002) A fast and elitist multiobjective genetic algorithm: Nsga-ii. Evol Comput, IEEE Trans 6(2):182–197

Deift P, Zhou X (1993) The steepest descent method for oscillatory Riemann–Hilbert problems. Asymptotics for the MKdV equation. Ann Math 137(2):295–368

Dennis JE Jr, Moré JJ (1977) Quasi-Newton methods, motivation, and theory. SIAM Review 19(1):46–89

Dorrington KP, Link CA (2004) Genetic-algorithm/neural-network approach to seismic attribute selection for well-log prediction. Geophysics 69(1):212–221

Du Z, MacGregor LM (2010) Reservoir characterization from joint inversion of marine CSEM and seismic AVA data using Genetic algorithms: a case study based on the Luva gas field. In: SEG technical program expanded abstracts, society of exploration geophysicists, pp 737–741

Geman S, Geman D (1984) Stochastic relaxation, Gibbs distributions, and the Bayesian restoration of images. IEEE transactions on pattern analysis and machine intelligence PAMI 6(6):721–741

Geman S, Geman D (1987) Stochastic relaxation, Gibbs distributions, and the Bayesian restoration of images. In: Readings in computer vision. Morgan Kaufmann, pp 564–584

Gill PE, Murray W, Wright MH (1981) Practical optimization. Academic Press, London

Goffe WL, Ferrier GD, Rogers J (1994) Global optimization of statistical functions with simulated annealing. J Econometrics 60(1–2):65–99

Goldberg DE, Holland JH (1988) Genetic algorithms and machine learning. Mach Learn 3(2):95–99

Hejazi F, Toloue I, Jaafar MS, Noorzaei J (2013) Optimization of earthquake energy dissipation system by genetic algorithm. Comput Aided Civ Inf 28(10):796–810

Holland J (1975) Adaptation in natural and artificial systems: an introductory analysis with application to biology. In: Control and artificial intelligence

Jervis M, Sen MK, Stoffa PL (1996) Prestack migration velocity estimation using nonlinear methods. Geophysics 61(1):138–150

Kelley CT (1999) Iterative methods for optimization. Soc Ind Appl Math

Kirkpatrick S, Gelatt CD, Vecchi MP (1983) Optimization by simulated annealing. Science 220(4598):671–680

Ma XQ (2002) Simultaneous inversion of pre-stack seismic data for rock properties using simulated annealing. Geophysics 67:1877–1885

Mallick S (1999) Some practical aspects of prestack waveform inversion using a genetic algorithm: an example from the east Texas Woodbine gas sand. Geophysics 64(2):326–336

Mallick S (1995) Model-based inversion of amplitude-variations-with-offset data using a genetic algorithm. Geophysics 60(4):939–954

Maurya SP, Singh KH, Kumar A, Singh NP (2017) Reservoir characterization using post-stack seismic inversion techniques based on a real coded genetic algorithm. J Geophys 39(2):95–103

Metropolis N, Rosenbluth AW, Rosenbluth MN, Teller AH, Teller E (1953) Equation of state calculations by fast computing machines. J Chem Phys 21(6):1087–1092

Misra S, Sacchi MD (2008) Global optimization with model-space preconditioning: application to AVO inversion. Geophysics 73(5):R71–R82

Moncayo E, Tchegliakova N, Montes L (2012) Pre-stack seismic inversion based on a genetic algorithm: a case from the Llanos Basin (Colombia) in the absence of well information. CT&F-Ciencia Tecnología y Futuro 4(5):5–20

Morgan EC, Vanneste M, Lecomte I, Baise LG, Longva O, McAdoo B (2012) Estimation of free gas saturation from seismic reflection surveys by the genetic algorithm inversion of a P-wave attenuation model. Geophysics 77(4):R175–R187

Mosegaard K, Vestergaard PD (1991) A simulated annealing approach to seismic model optimization with sparse prior information 1. Geophys Prosp 39(5):599–611

Polyak BT (1969) The conjugate gradient method in extremal problems. USSR Comput Math Math Phys 9(4):94–112

Pullammanappallil SK, Louie JN (1994) A generalized simulated-annealing optimization for inversion of first-arrival times. Bull Seismol Soc Am 84(5):1397–1409

Rebbi C (1987) Monte Carlo calculations in lattice gauge theories. Applications of the Monte Carlo method in statistical physics. Springer, Berlin, pp 277–297

Sen MK, Stoffa PL (1991) Nonlinear one-dimensional seismic waveform inversion using simulated annealing. Geophysics 56(10):1624–1638

Sen MK, Stoffa PL (2013) Global optimization methods in geophysical inversion. Cambridge University Press, Cambridge

Sen MK, Datta-Gupta A, Stoffa PL, Lake LW, Pope GA (1995) Stochastic reservoir modeling using simulated annealing and genetic algorithm. SPE Formation Eval 10(01):49–56

Shewchuk JR (1994) An introduction to the conjugate gradient method without the agonizing pain, Technical report, Carnegie Mellon University

Sivanandam SN, Deepa SN (2007) Introduction to genetic algorithms. Springer Science and Business Media

Szu H, Hartley R (1987) Fast simulated annealing. Phys Lett A 122(3–4):157–162

Tran KT, Hiltunen DR (2012) One-dimensional inversion of full waveforms using a genetic algorithm. J Environ Eng Geophys 17(4):197–213

Van Laarhoven PJ, Aarts EH (1987) Simulated annealing. Simulated annealing: theory and applications. Springer, Dordrecht, pp 7–15

Velez-Langs O (2005) Genetic algorithms in the oil industry: an overview. J Petrol Sci Eng 47(1):15–22

Whitley LD (1989) The GENITOR algorithm and selection pressure: why rank-based allocation of reproductive trials is best. Icga 89:116–123

Yang J, Honavar V (1998) Feature subset selection using a genetic algorithm. In: Feature extraction, construction and selection. Springer, Berlin, pp 117–136

Yan-Jun H, Ding-Hui Y, Yuan-Feng C (2016) Reservoir parameter inversion of co2 geological sequestration based on the self-adaptive hybrid genetic algorithm. Chin J Geophys Chin Ed 59(11):4234–4245

Yuan Y, Zhuang H (1996) A genetic algorithm for generating fuzzy classification rules. Fuzzy Sets Syst 84(1):1–19

Chapter 7
Geostatistical Inversion

Abstract Geostatistical inversion methods are routinely used to predict various geophysical parameters away from the boreholes using seismic and well log data. The geostatistics derives a surface using the values from the measured locations to estimate data points for each location in between the data points. The present chapter discusses different types of seismic attributes and their use in the interpretation of seismic data. Thereafter, four types of geostatistical methods namely single attribute analysis, multi-attribute regression, probabilistic neural network, and multi-layer feed-forward neural network methods are discussed. Initially, the mathematical background of these methods has been discussed and finally, the application of these methods to the real data is provided for better understanding.

7.1 Introduction

Geostatistical inversions are a mathematical technique that integrates seismic and well logs data to predict various rock properties in the inter-well region. The techniques used to predict parameter values between information points. Geostatistical inversions are valid only for information relating to space (i.e., non-random). The fundamental of technique is first to define and quantify the spatial structure of the variables of interest and then to interpolate or assess the values of adjacent values based on their spatial composition. There are many types of geostatistical inversion techniques are available and continuously developing but some of them are very important and routinely used by the petroleum industry. These are single attribute analysis, multi-attribute regression and neural network which include feed-forward neural network and probabilistic neural network. These methods use seismic inversion derived impedance as external attributes and attributes that are derived directly from the seismic data as internal attributes for its implementation (Doyen 1988). These attributes are analyzed among themselves and the combinations of best attributes are selected to derive the relationship between target logs (petrophysical property to be predicted) and attributes at each sample point which is further used to predict well log property in the inter-well region (Haas and Dubrule 1994).

S. P. Maurya et al., *Seismic Inversion Methods: A Practical Approach*, Springer Geophysics, https://doi.org/10.1007/978-3-030-45662-7_7

Two groups of interpolation techniques are provided by Geostatistics namely deterministic and geostatistical (Russell et al. 1997). Mathematical functions are used by the Deterministic techniques for interpolation whereas Geostatistics uses both statistical and mathematical methods (Hampson et al. 2001).

The geostatistical technique, with its origins in the hard-rock mining industry, is very useful for reservoir characterization in the exploration geophysics (Bosch et al. 2010). The initial use of geostatistical methods was limited due to dependence on the requirement for well log density but modern methods, however, employ a successful integration of well logs (high frequency) and seismic information (low frequency) and hence their area of the application becomes wide-ranging. The inputs to geostatistical inversion no longer consist of traditional seismic reflection data (Lindseth 1979).

These methods can be used to predict velocity, density, gamma-ray, porosity, permeability, shale/sand ratio, etc. but in most case porosity and permeability is the major petrophysical parameters needs to analyze for reservoir characterization. The difficulty of estimating porosity and permeability due to its wide variation over a reservoir volume, but are only be sampled at well locations, using different technologies at various scales of observation (Chiles 1988; Taner et al. 1979). Seismic data are mostly affected by porosity and rock properties of the subsurface. In some cases, seismic data may also be affected by pore fluid replacement, which is generally controlled by permeability (Todorov 2000). Thus an accurate prediction of porosity and permeability is of great importance in the exploration project (Jones and Helwick 1998).

The quantitative study of the porosity of a rock is also very important but it is not too easy. The main problem to estimate porosity is the presence of several types of porosity of the formation (Hampson et al. 2001). Porosity can be estimated by various techniques, some techniques are based on core samples; other techniques are used well-log data through mathematical models (Dubrule 2003). Well log data provide very good vertical resolution of the subsurface (geologic strata), but generally, they are at sparse locations (Adeli and Panakkat 2009). In contrast, seismic data provides very dense and regular areal sampling but the vertical resolution is lower. The integration of seismic data with wells can therefore significantly improve the prediction of porosity (Masri et al. 2000). The use of seismic attributes is a widely used method to decrease the spatial uncertainty of the prediction (Specht 1990, 1991).

We attempt to go beyond the limits of conventional seismic inversion in several ways which are already discussed in earlier chapters. Firstly, predict log properties other than acoustic impedance, such as porosity (Anderson 1996). Secondly, use attributes derived from the seismic data rather than the conventional post-stack data. This allows including pre-stack information as well as nonlinear transforms of the post-stack data. Thirdly, instead of assuming a particular model relating the logs and the seismic data, a statistical relationship will be derived by analyzing a set of training data at well locations. This relationship will be either linear (multivariate regression) or nonlinear (neural network). Finally, we will use the concept of cross-validation to estimate the reliability of the derived relationship (Torres-Verdin et al. 1999). A flowchart of geostatistical inversions is presented in Fig. 7.1. The figure demonstrates the working principle of all geostatistical methods discussed in this

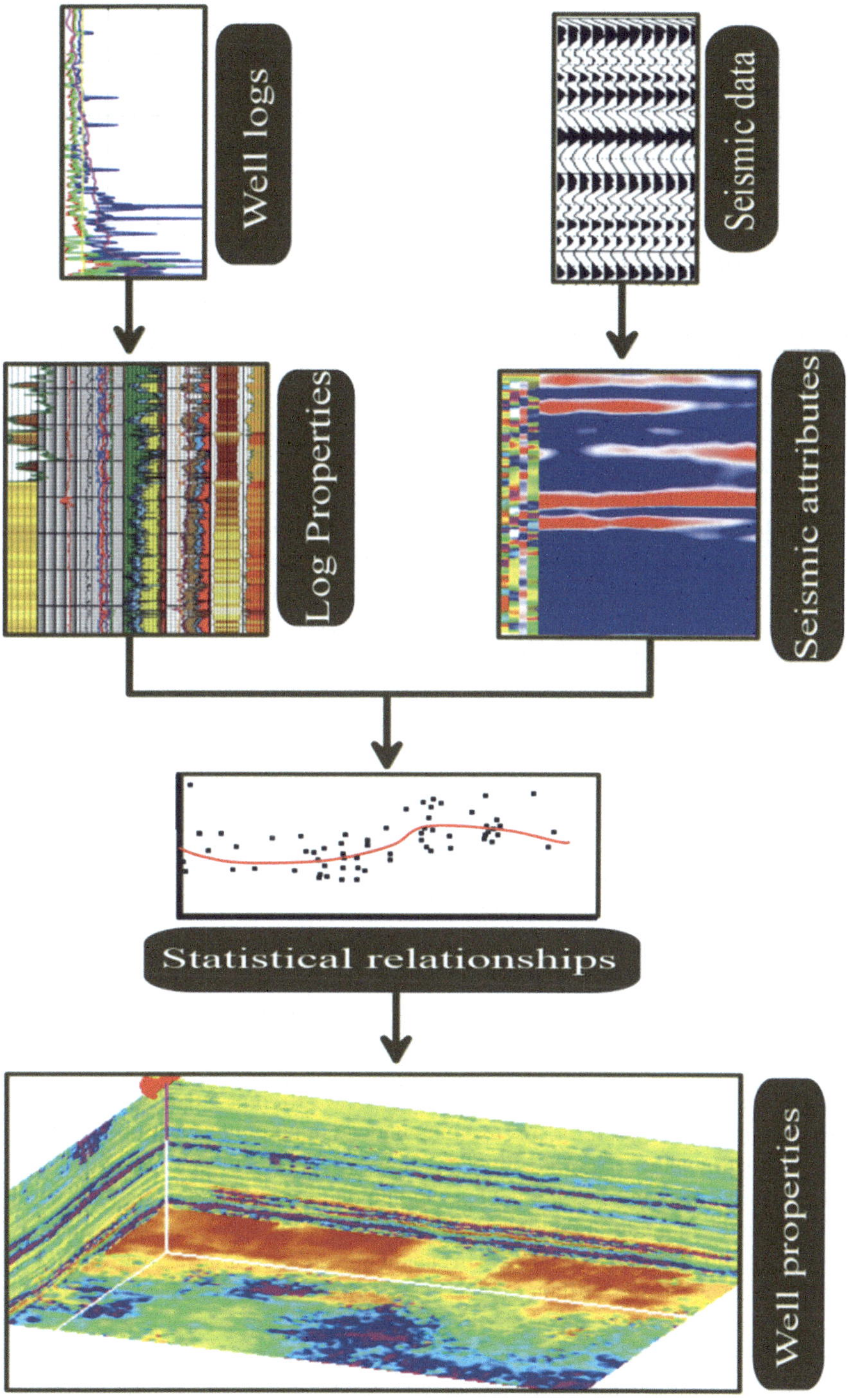

Fig. 7.1 Flowchart of geostatistical methods

chapter. The geostatistical inversions utilize a number of seismic attributes estimated directly or indirectly from the seismic reflection data and hence their understanding becomes very important before moving towards geostatistical inversions in detail. The following sections are discussed some important seismic attributes and their uses.

7.2 Seismic Attributes

Seismic attributes can be conveniently defined as the quantities that are measured, computed or implied from the seismic data. From the time of their introduction in the early 1970s, seismic attributes have gone along the way and they became an aid for geoscientists for reservoir characterization and also as a tool for quality control (Chambers and Yarus 2002; Chopra and Marfurt 2007). Seismic attributes are specific measurements of geometric, kinematic, dynamic, or statistical features derived from seismic data. Some are more sensitive than others to specific reservoir environments; some are better at revealing subsurface anomalies that are not easily detectable, and some have been used as direct hydrocarbon indicators (Taner et al. 1994; Chopra and Marfurt 2005, 2007). Different authors introduced different kinds of attributes and their uses. With the introduction of 3D seismic techniques and associated technologies and introduction of seismic sequence attributes, coherence technology in the mid-1990s, and spectral decomposition in the late 1990s has changed the seismic interpretation techniques and provided essential tools that were not available for geoscientists earlier. With the introduction of 3D visualization techniques, the use of seismic attributes has attained a new dimension. The development of a wide variety of seismic attributes warrants a systematic classification. Also, a systematic approach is needed to understand the use of each of these attributes and also their limitations under different circumstances.

7.2.1 Classification of Seismic Attributes

Though the purpose of this book is to understand different attributes that can be used as tools in interpretation, it is useful to understand the classification of different attributes at this stage. The more seismic attributes available, the more confusion geoscientists may have in selecting the appropriate ones. Improved documentation and a greater understanding of the underlying rock physics that give rise to an anomaly, the role that seismic processing plays in enhancing or destroying attribute anomalies, and the value of seismic attributes in the absence of good well control are areas where assistance is needed (Russell et al. 1997). Guidance should be available to the interpreter to help with the choice of the seismic attributes that will provide the greatest insight into the current problem under investigation. The one way to classify seismic attributes is based on their characteristics and hence geometrical and physical.

7.2.1.1 Geometrical Attributes

The attributes representing an object's (polygon segment) geometric features create an object-based image analysis that is termed as a geometrical attribute. Geometric attributes describe all the other attributes spatial and temporal relationships. A useful marker of bedding resemblance as well as discontinuity is the lateral continuity measured by semblance. Depositional data is provided by bedding dips and curvatures (Brown 1996). Geometric attributes are also useful for stratigraphic analysis as they identify the characteristics of events and their spatial relationships and can be used to quantify characteristics that directly help in the recognition of deposition patterns and associated lithology.

The Geometrical attributes are dip, azimuth, and discontinuity. The Dip attribute or amplitude of the data corresponds to the dip of the seismic events. The dip is useful in that it makes faults more discernible. The amplitude of the data on the Azimuth attribute corresponds to the azimuth of the maximum dip direction of the seismic feature (Taner 2001).

7.2.1.2 Physical Attributes

Physical attributes are defined as those attributes which are directly related to the wave propagation, lithology, and other parameters. These physical attributes can be further classified as pre-stack and post-stack attributes. Each of these has sub-classes as instantaneous and wavelet attributes. Instantaneous attributes are computed sample by sample and indicate a continuous change of attributes along the time and space axis (Brown 1996; Taner 2001; Maurya et al. 2019). The Wavelet attributes, on the other hand, represent characteristics of wavelet and their amplitude spectrum. The magnitude of the trace envelope is proportional to the contrast between the acoustic impedance; frequencies are related to the density of the bed, wave dispersion, and absorption. Consequently, these attributes are mostly used for lithological classification and reservoir characterization. The classification of physical attributes is as follows.

(i) Post stack attributes
(ii) Pre-stack attributes

i. Post-stack attributes

Post stack attributes are derived from the stacked seismic data. Stacking is an average method that removes data linked to offset and azimuth. It may be possible to stack or migrate CDP input information. It should be noted that time migrated information will retain their time relationships, thus temporal factors such as frequency will retain their physical characteristics as well. Frequency is substituted by wave number for depth migrated data, which is a function of velocity and frequency of propagation. Post-stack characteristics are a more manageable strategy in original recognition inquiries for looking big quantities of information.

The Attributes are the result of the properties derived from the complex seismic signal. The concept of complex traces was first described by Taner et al. (1979). The complex trace is defined as follows.

$$CT(t) = T(t) + iH(t) \tag{7.1}$$

where $CT(t)$ represent complex trace, $T(t)$ represent seismic trace, $H(t)$ is Hilbert's transform of a seismic trace (T(t)). Further, post-stack attributes are classified into many subparts based on complex trace and some of them are discussed in the following sections.

(a) Signal envelope (E)

The Signal envelope (E) is also called Reflection Strength is calculated from the complex trace as follows.

$$E(t) = SQRT\left(T^2(t) + H^2(t)\right) \tag{7.2}$$

where SQRT represents square root. The signal envelope of the seismic signal has a low-frequency appearance with only positive amplitudes. It often highlights the main seismic features. The envelope represents the instantaneous energy of the signal and is proportional in its magnitude to the reflection coefficient (Brown 1996).

The envelope is useful in highlighting discontinuities, changes in lithology, faults, changes in deposition, tuning effect, and sequence boundaries. It also is proportional to reflectivity and therefore useful for analyzing AVO anomalies. If there are two volumes that differ by constant phase shift only, their envelopes will be the same (Russell et al. 2001). This attribute represents mainly the acoustic impedance contrast, hence reflectivity. This attribute is mainly useful in identifying bright spots, gas accumulation, Sequence boundaries, major changes or depositional environments, Thin-bed tuning effects, Unconformities, Major changes of lithology.

(b) Envelope derivative (RE)

Envelope derivative or time rate of envelope shift indicates the energy variability of the reflected events. Events with a sharp relative increase also indicate a broader bandwidth and therefore fewer impacts on absorption (Chopra and Marfurt 2007). This attribute is also a physical characteristic and can be used to identify potential impacts of fracturing and absorption. The envelope derivative can be written mathematically as follows.

$$RE(t) = \frac{dE(t)}{dt} \tag{7.3}$$

The derivative of the envelope highlights the

- Change in reflectivity and is also related to the absorption of energy.
- The sharpness of the rise time relates to absorption sharp interfaces
- Shows discontinuities,

It is used in the computation of group propagation direction. When compared with the phase propagation direction, it may indicate dispersive waves.

(c) Second derivative of envelope (DDE)

The second derivative of the envelope provides a measure of the maximum envelope's sharpness. It can be used to define all seismic bandwidth reflecting interfaces (Chopra and Marfurt 2005). This is given as follows.

$$DDE(t) = \frac{d^2E(t)}{dt^2} \tag{7.4}$$

The second derivative of the envelope highlights the interfaces very well. This attribute is not too sensitive to the amplitude and can highlight even weak events. It is used to show all reflecting interfaces visible within seismic bandwidth, the sharpness of events, sharp changes of lithology (Chopra and Marfurt 2007). It is a very good presentation of the image of the subsurface within the seismic bandwidth.

(d) Instantaneous phase

Due to the definition of wavefronts as lines of the continuous stage, the phase attribute is also a physical characteristic and can be used efficiently as a discriminator for the classification of geometric shapes. The instantaneous phase attribute is given as follows.

$$\phi(t) = \text{arc}\,tan\left|\frac{H(t)}{T(t)}\right| \tag{7.5}$$

The seismic trace $T(t)$ and its Hilbert transform $H(t)$ are related to the envelope $E(t)$ and the phase $\phi(t)$ by the following relation.

$$T(t) = E(t)\cos(\phi(t))$$

$$H(t) = E(t)\sin(\phi(t))$$

The instantaneous phase is measured in degrees $(-\pi, \pi)$. It is independent of amplitude and shows continuity and discontinuity of events (Chambers and Yarus 2002). It shows bedding very well. This attribute is useful as the best indicator of lateral continuity, relates to the phase component of the wave- propagation, can be used to compute the phase velocity, shows discontinuities, but may not be the best, has no amplitude information, hence all events are represented. It is better to show continuities sequence boundaries, used in the computation of instantaneous frequency and acceleration.

(e) Cosine of instantaneous phase C(t)

The Cosine of Instantaneous Phase can be mathematically estimated as follows.

$$C(t) = \cos(\phi(t)) \tag{7.6}$$

The cosine of the instantaneous phase is also independent of amplitude and shows bedding very well. This attribute is smoother than phase (which has discontinuities) and thus is useful for automatic classification procedures.

(f) Instantaneous acceleration AC(t)

Instantaneous acceleration is the time derivative of the instantaneous frequency. It highlights the change in instantaneous frequency which may be indicative of thin bedding or absorption. Mathematically, the AC can be represented as follows.

$$AC(t) = \frac{d(F(t))}{dt} \tag{7.7}$$

The instantaneous acceleration used to accentuates bedding differences, higher resolution, may have a somewhat higher noise level due to differentiation, may have some relation to elastic properties of beds.

(g) Instantaneous frequency (IF)

Instantaneous frequency is the time derivative of the phase, i.e., the rate of change of the phase. The instantaneous frequency attribute has been shown to be related to the centroid of the seismic wavelet power spectrum (Cohen and Lee 1990; Barnes 1991, 1993). The instantaneous frequency attribute reacts to both impacts of wave propagation and deposition features, so it is a physical attribute and can be used as an efficient discriminator. Mathematically, it can be written as follows.

$$IF(t) = \frac{d(\phi(t))}{dt} \tag{7.8}$$

Instantaneous frequency represents the mean amplitude of the wavelet. Instantaneous frequency can indicate bed thickness and also lithology parameters. It indicates the edges of low impedance thin beds and also used for hydrocarbon indicator and fracture zone indicator by low-frequency anomaly. This effect is sometimes accentuated by the unconsolidated sands due to the oil content of the pores.

(h) Thin bed indicator

This attribute is designed to highlight the location where the instantaneous frequency jumps or goes in the reverse direction (Tanner 1979). These jumps could be the result of very close reflectors, i.e., thin beds. It is calculated as follows.

$$TB(t) = F(t) - F'(t) \tag{7.9}$$

For this attribute, calculation window length is used to calculate the $F'(t)$. It computed from large spikes of instantaneous frequency, indicate overlapped events, indicates thin beds when laterally continuous indicates non-reflecting zone when it appears laterally at random, like salt and pepper and also used to fined detail of bedding patterns.

(i) Instantaneous bandwidth B(t)

This attribute is described by Barnes 1991. It can be written as follows.

$$B(t) = \frac{\frac{d(E(t))}{dt}}{2.\mathrm{E(t)}} \tag{7.10}$$

- This attribute highlights bedding very well.
- Represents the seismic data bandwidth sample by sample. It is one of the high-resolution character correlators.
- Shows the overall effects of absorption and seismic character changes (Russell et al. 2001).

(j) Instantaneous Q

Instantaneous Q is an attribute suggested by Barnes (1992). Q is the quality factor that is related to attenuation. Instantaneous Q measures the high-frequency component of Q and shows local variations of Q. This attribute is derived as follows.

$$Q(t) = *IF(t) * \frac{E(t)}{\frac{\partial E(t)}{\partial t}} \tag{7.11}$$

It used to indicate local variation of Q factor, similar to the relative acoustic impedance computation from the seismic trace. Longer wavelength variation should be computed by spectral division and added to this attribute. It may indicate liquid content by rationing pressure versus shear wave section Q factors. It also indicates the relative absorption characteristics of beds (Taner 2001).

(k) Relative acoustic impedance

This attribute calculates the running sum of the trace to which a low cut filter is applied. It is an indicator of impedance changes, in a relative sense. The low cut filter is applied to remove DC shift which is typical in impedance data. (If the value of the low cut filter is zero, then it is not applied). The calculated trace is the result of simple integration of the complex trace (Maurya and Singh 2018). It represents the approximation of the high-frequency component of the relative acoustic impedance.

ii. Pre-stack attributes

Pre-stack attributes are calculated from NMO corrected, migrated pre-stack seismic data. They will have data linked to a direction (azimuth) and offset. The pre-stack seismic data have more information and hence the pre-stack attributes are used to estimate more properties as compared with the post-stack attributes (Taner 2001; Barnes 1994; Maurya and Singh 2018). These calculations produce enormous quantities of data; therefore they are not practical for original research. They contain significant quantities of data, however, which can be directly connected to fluid content and fracture orientation. This class includes AVO, velocities and azimuthal variety of all characteristics. The pre-stack attributes are further classified.

(a) RMS velocities of reflectors

This may be calculated from time migration velocity analysis, independent of the major influence of dips. This is used for sand/shale ratios estimation, high-pressure shale zone detection, major lithologic change detection, and etc.

(b) Discontinuity

It is again a geometrical attribute and measures the lateral relations in the data. It is designed to emphasize the discontinuous events such as faults. High amplitude values on this attribute correspond to discontinuities in the data, while low amplitude values correspond to continuous features. Discontinuity varies between zero and one, where zero is continuous and one is discontinuous. The derivation of the attribute is prepared by considering several traces together to reveal the discontinuity (geometry) of the beds (Soubotcheva and Stewart 2004). This attribute can be used to understand, coherency at maximum coherency direction, minimum coherency direction, event terminations, picked horizons, fault detection, zones of parallel bedding, zones of chaotic bedding, non-reflecting zones, converging and diverging bedding patterns and unconformities.

Some important attributes are estimated from the Blackfoot seismic data and presented in Figs. 7.2 and 7.3. The high amplitude anomaly is highlighted by the arrow. One can also notice that some of the above-discussed attributes are dramatically increased visibility of subsurface features and hence seismic amplitudes are very important steps used in support of seismic data interpretation. Figures 7.4 and 7.5 shows slice at a 1060 ms time interval of all the attributes presented in Figs. 7.2 and 7.3. These slices represent the horizontal distribution of the attributes and also one can notice the special features of the subsurface are increased.

7.3 Single Attribute Analysis (SAA)

This is the first type of geostatistical inversion method used by geoscientists to predict subsurface rock properties away from the boreholes. In this methodology, the goal is to discover an operator potentially linear that can predict well logs from nearby information. Indeed, we choose not to evaluate the seismic information itself, but the seismic attributes. The reason behind is that many of these attributes will be nonlinear, thereby improving the technique's predictive power as compared with the raw seismic data. One other reason is that the breakdown of input information into a component is often beneficial. This method is called preprocessing or extraction of features and can often significantly enhance the efficiency of a pattern recognition scheme by decreasing information dimensionality (Russell et al. 2001).

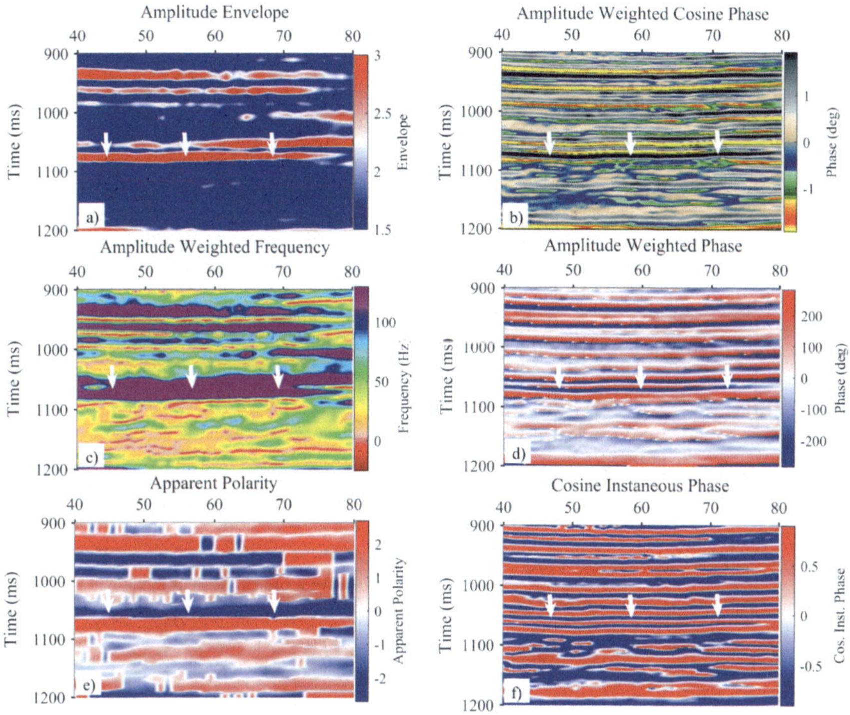

Fig. 7.2 Shows seismic attributes estimated from the Blackfoot seismic data sets, where **a** represents amplitude envelope, **b** amplitude weighted cosine phase, **c** amplitude weighted frequency, **d** amplitude weighted phase, **e** apparent polarity and **f** represents cosine instantaneous phase

The single attribute analyzes all or some of the inner and external attributes to determine the most helpful attribute for anticipating target logs. The single attribute expression implies the attributes are used separately as opposed to the Multi-attribute where the characteristics are used in groups.

Single attribute analysis processes in which one attribute at a time used to find out the relation between attribute and desired well log property. In the first process toward prediction of petrophysical parameters, seismic attributes are extracted from the seismic data. In the second step, the best attribute is selected. Thereafter, that particular attribute is cross plotted with the target log for deriving a linear relationship that is utilized to predict log property.

Table 7.1 contains a list of attributes and their correlation with desired well log property. Each line of the table contains information for a particular attribute. Usually, it is best to use all the attributes and rank them according to their accuracy to predict the target logs. The attributes are ordered by increasing prediction error, i.e., by decreasing suitability for predicting the target. Nonlinear transformations will check to see if the inverse, log, square root or other easy transformations will generate a linear connection between good log and seismic attributes. For example,

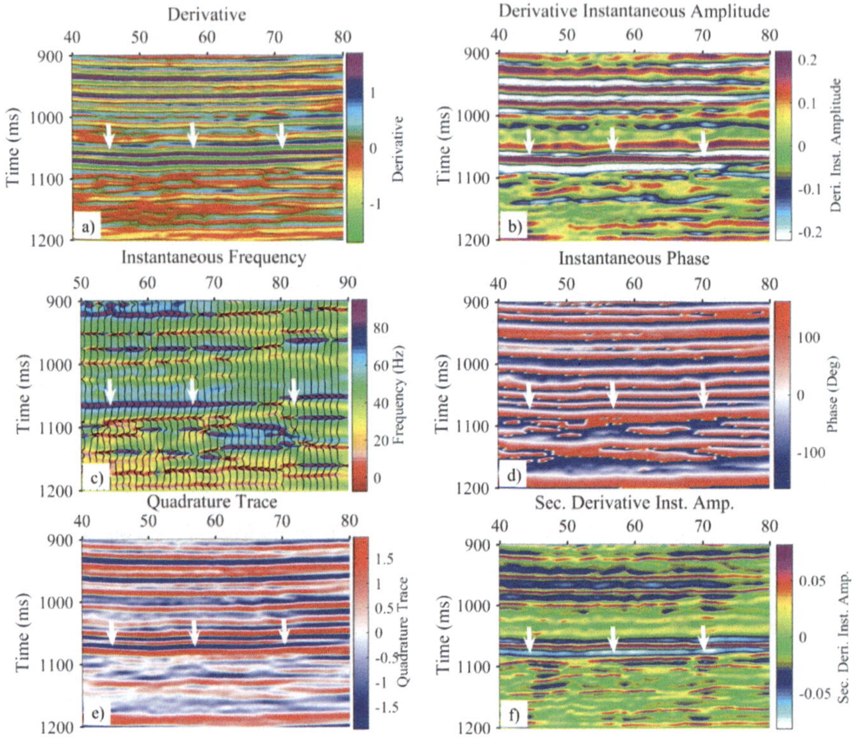

Fig. 7.3 Shows seismic attributes estimated from the Blackfoot seismic data sets, where **a** represents derivative, **b** derivative instantaneous amplitude, **c** instantaneous frequency, **d** instantaneous phase, **e** quadrature trace and **f** second derivative instantaneous amplitude

a relationship between a log attribute and a seismic attribute may not be linear (it is cross plot may resemble a curve), but if one or both of the attributes are transformed into log values, a linear relationship may appear, with the points showing a straight line on the cross plot.

From Table 7.1, one can notice that attribute $1/impedance$ is the best one and hence can be used for the prediction process. Thereafter, this selected attribute is cross-plotted with well log porosity (density porosity) to predict porosity in the inter-well region and represented in Fig. 7.6. In this analysis, the assumption is that the target log has been integrated to travel time at the same sample rate as the seismic attribute. Effectively, this integration reduces the target log to the same resolution as the attribute, which is usually significantly coarser than the log property. Each point in the cross plot consists of a pair of numbers corresponding to a particular time sample.

Assuming a linear relationship between the target log (density porosity) and the attribute, a straight line may be fit by regression in the following way.

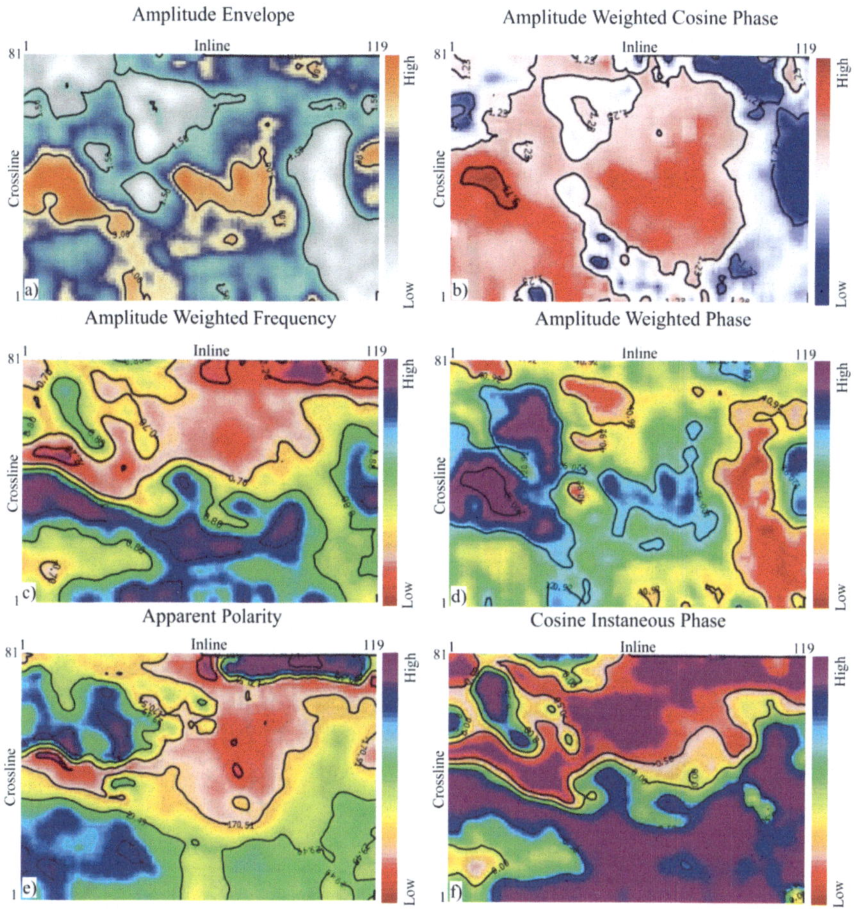

Fig. 7.4 Shows slices at 1065 ms time from seismic attributes estimated from the Blackfoot seismic data sets, where **a** represents amplitude envelope, **b** amplitude weighted cosine phase, **c** amplitude weighted frequency, **d** amplitude weighted phase, **e** apparent polarity and **f** represents cosine instantaneous phase

$$y = qx + p \tag{7.12}$$

The coefficients p and q in the above equation may be derived by minimizing the mean-squared prediction error as following.

$$E^2 = \frac{1}{N}\sum_{i=1}^{N}(y_i - p - qx_i)^2 \tag{7.13}$$

where the sum is overall points in the cross plot. The calculated prediction error E is a measure of the goodness-of-fit for the regression line defined by Eq. (7.12). An

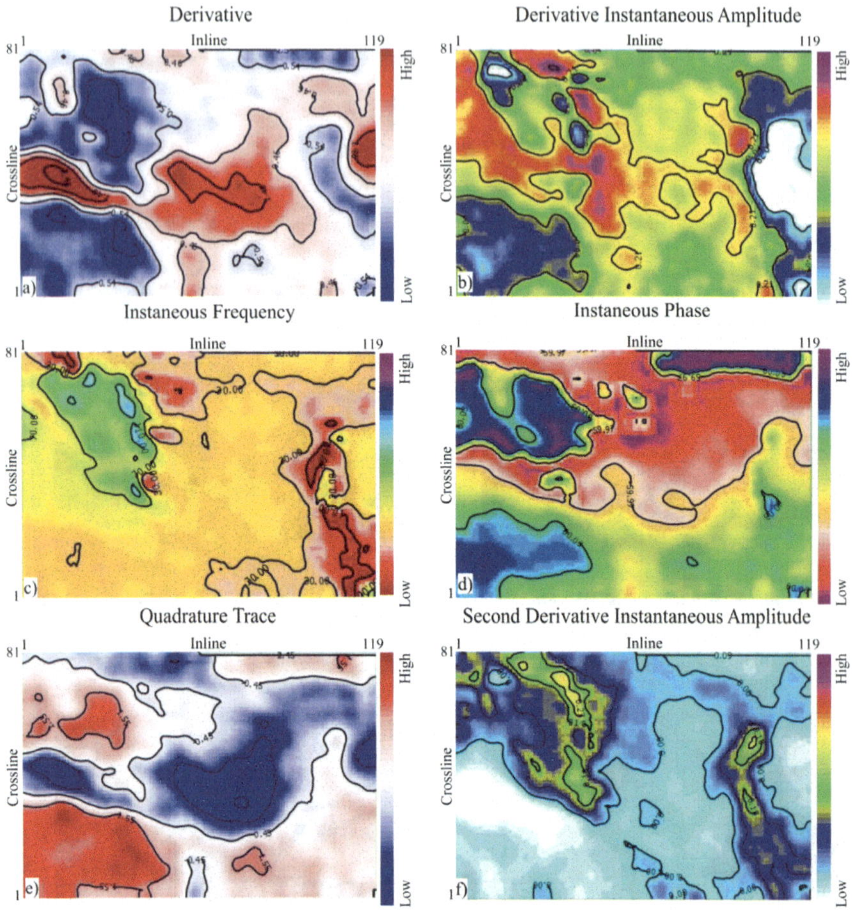

Fig. 7.5 Shows slices at 1065 ms time of seismic attributes estimated from the Blackfoot seismic data sets, where **a** represents derivative, **b** derivative instantaneous amplitude, **c** instantaneous frequency, **d** instantaneous phase, **e** quadrature trace and **f** second derivative instantaneous amplitude

alternative measure is the normalized correlation coefficient, defined as follows.

$$\rho = \frac{\sigma_{xy}}{\sigma_x \sigma_y} \tag{7.14}$$

where,

$$\sigma_{xy} = \frac{1}{N} \sum_{i=1}^{N} (x_i - m_x)(y_i - m_y) \tag{7.15}$$

Table 7.1 Single attribute list and their correlation with target porosity

S. No.	Target	Attribute	Error	Correlation
1.	Porosity	1/(Impedance)	5.01322	0.381154
2.	Porosity	Impedance	5.041142	−0.36842
3.	Porosity	(Impedance) * 2	5.066357	−0.35646
4.	Porosity	Derivative	5.261693	0.24177
5.	Porosity	Quadrature trace	5.275328	−0.23145
6.	Porosity	Amplitude weighted cosine phase	5.275942	0.230968
7.	Porosity	Filter	5.306574	0.205723
8.	Sqrt (porosity)	Derivative	5.308363	0.249109
9.	Porosity	Amplitude weighted phase	5.320532	−0.19307
10.	Porosity	Cosine instantaneous phase	5.343901	0.169711
11.	Porosity	Instantaneous phase	5.36141	−0.14976
12.	Porosity	Derivative instantaneous amplitude	5.400413	0.090291
13.	Porosity	Second derivative	5.403192	−0.08445
14.	Porosity	Integrated absolute amplitude	5.403278	−0.08426
15.	Porosity	Instantaneous frequency	5.411403	0.064118

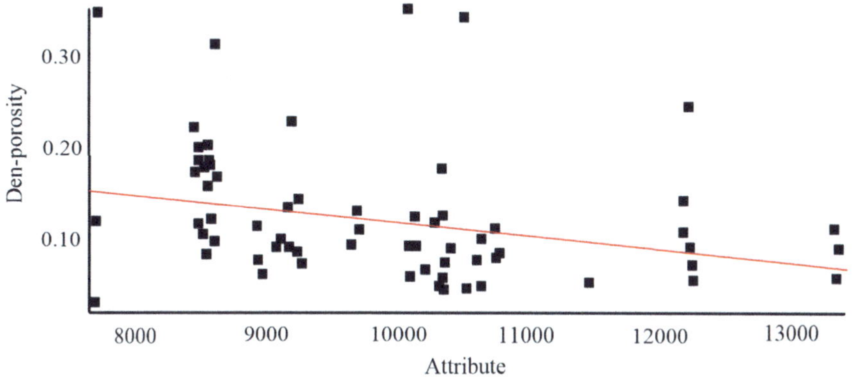

Fig. 7.6 Crossplot of density porosity with attributes. A best fit straight line is fitted to the data (redline)

$$\sigma_x = \frac{1}{N}\sum_{i=1}^{N}(x_i - m_x) \tag{7.16}$$

$$\sigma_y = \frac{1}{N}\sum_{i=1}^{N}(x_i - m_y) \tag{7.17}$$

$$m_x = \frac{1}{N}\sum_{i=1}^{N}(x_i) \tag{7.18}$$

$$m_y = \frac{1}{N}\sum_{i=1}^{N}(y_i) \tag{7.19}$$

Numerical relationship is set up between attributes and used to predict den-porosity at well locations.

Figure 7.7 shows a comparison of target logs in black with the predicted logs (estimated using SAA) in red solid line for six wells. The Average error reported at the top of the plot is the root-mean-square difference between the target log values and the predicted values, calculated over the analysis window. Notice that applying a regression curve to the 1/(*Impedance*) (Table 7.1) produces a result that tracks the general trend of the target logs, but does not adequately predict the subtle features. This is because the inversion result has been blocked with a relatively coarse block size.

It is noticed from Fig. 7.7 that the predicted porosity is matching with actual porosity in some extend. The statistical parameter i.e. the average correlation coefficient is estimated to be 0.45 and the average RMS error is 4.85. A cross plot of the predicted porosity with the actual porosity is also generated and shown in Fig. 7.8. The distribution of scatter points demonstrates that the predicted porosity is following the trend of actual porosity. The cross-section of predicted porosity at inline 27 is shown in Fig. 7.9. The vertical, as well as the horizontal variation of the porosity (0–15%), can be noticed from these sections. The 3D volume of predicted porosity is presented in Fig. 7.10. The section shows a smooth variation of porosity and hence

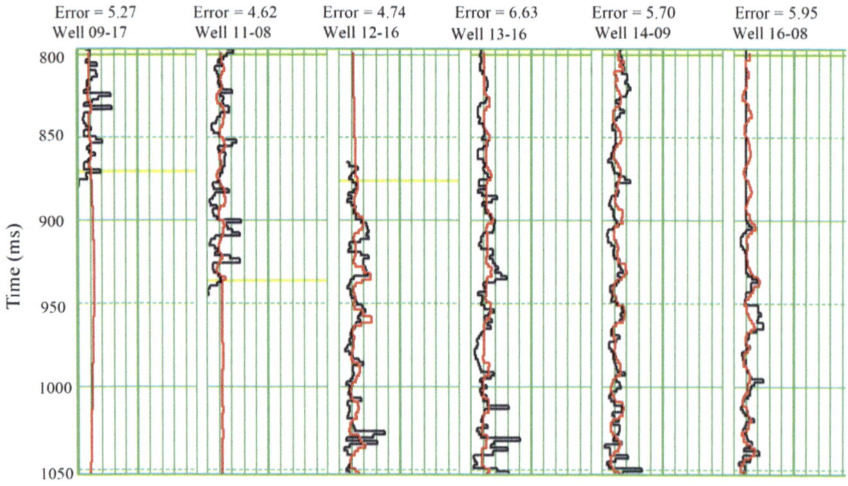

Fig. 7.7 The plot shows the target logs in black with the predicted logs in red. The average error and well name reported at the top of the plot

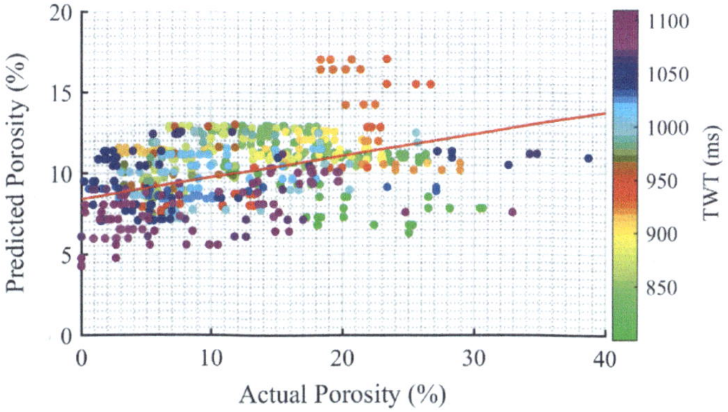

Fig. 7.8 Crossplot of modeled and actual porosity

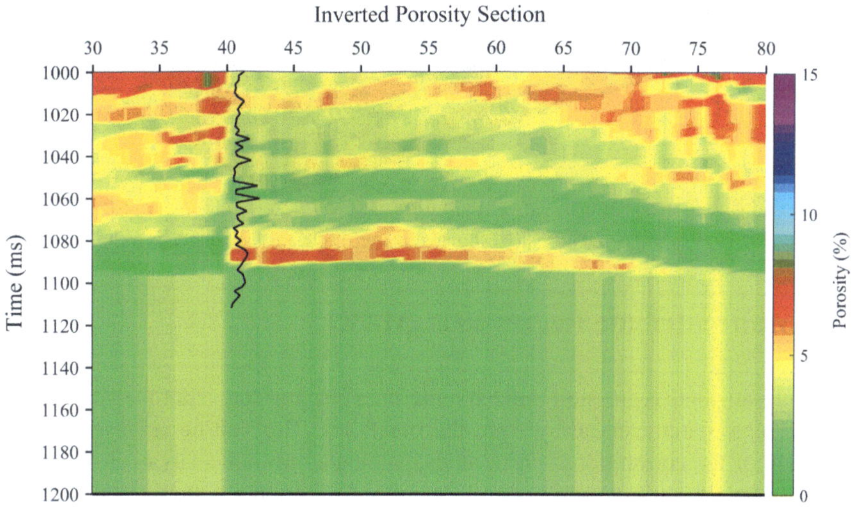

Fig. 7.9 Cross-section (inline 244) of predicted porosity estimated using a single attribute analysis algorithm

one can not differentiate the thin subsurface layer. This is because the prediction depends only on one attributes hence the resolution is poor.

The use of single attribute analysis over 3D datasets is very limited. In most cases, the single attribute analysis shows poor correlation of predicted curve with the actual curve and hence cannot use for quantitative interpretation. Although, the methods can be utilized for qualitative interpretation as the method is very fast and hence within a limited time, one can generate a volume of petrophysical properties. To overcome the problem encounter in single attribute analysis, the multi-attribute regression needs to be analyzed.

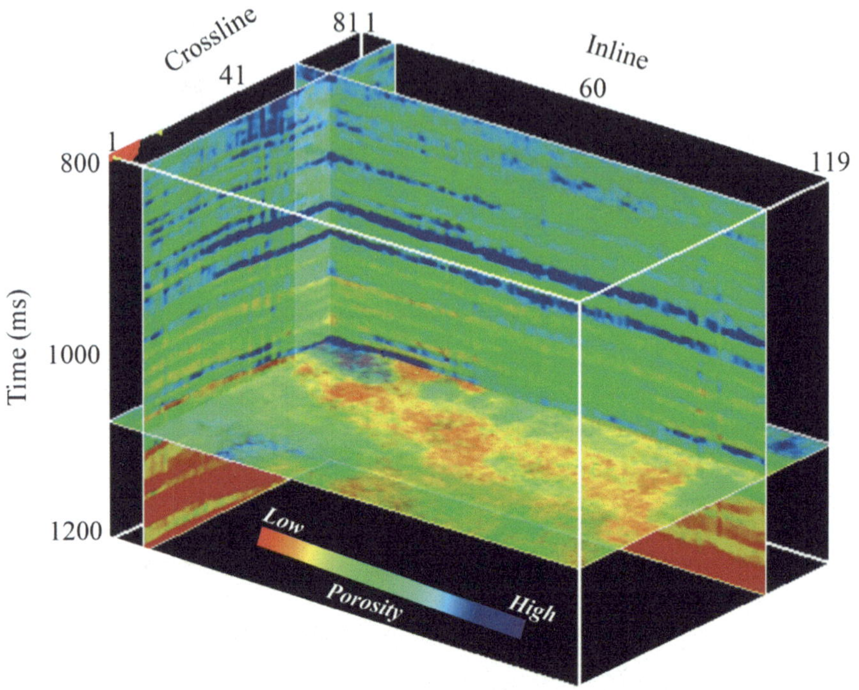

Fig. 7.10 3D view of predicted porosity estimated using single attribute analysis technique

7.4 Multi-attribute Regressions (MAR)

The multi-attribute regression utilized more than one attribute at a time and predict petrophysical parameters away from the boreholes. This can be achieved by the extension of the conventional linear analysis (single attribute analysis) to multiple attributes (multi-attribute regression). Assume, for simplicity, that we have three attributes (a_1, a_2 and a_3) good enough to predict porosity. Figure 7.11 shows the target log along with three attributes and represents how these attributes used to predict log value.

At each time sample, the target log is modeled by the linear equation as follows.

$$L(t) = w_0 + w_1 a_1(t) + w_2 a_2(t) + w_3 a_3(t) \tag{7.20}$$

where w_0, w_1, w_2, and w_4 are weights. The weights in Eq. 7.20 may be derived by minimizing the mean-squared prediction (Hampson et al. 2000, 2001) error, as follows.

$$E^2 = \frac{1}{N}\sum_{i=1}^{N}(L_i - w_0 - w_1 a_{1i} - w_2 a_{2i} - w_3 a_{3i})^2 \tag{7.21}$$

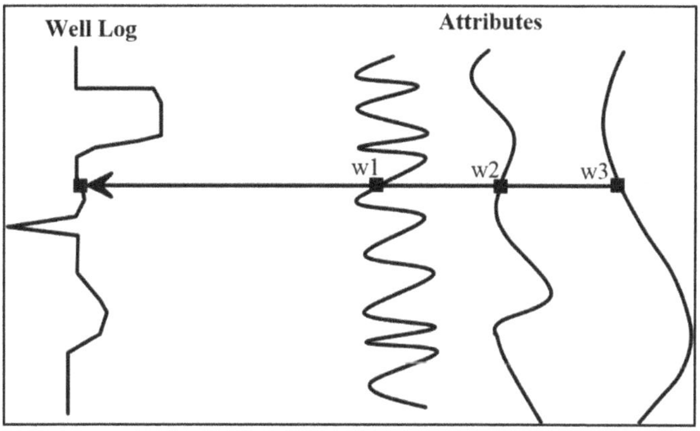

Fig. 7.11 Assuming the case of three seismic attributes, each target log sample is modeled as a linear combination of attribute samples at the same time

The solution for the four weights produces the standard normal equations which can be estimated in the following way.

Multi-attribute linear regression is an extension of simple attribute regression methods with M number of variables. Now we will use a M number of seismic attributes, $a_1, a_2, \ldots a_M$, to predict the log properties (L). To do this, one must determine the $M + 1$ weights $w_1, w_2, \ldots w_M$, which, when multiplied by the particular set of attribute values, give the closest result to the log in a least-squared sense. For simplicity, assume that $M = 3$ (Hampson et al. 2001). If one has N samples in log data, then, the following equations can be written.

$$
\begin{aligned}
L_1 &= w_0 + w_1 a_{11} + w_2 a_{21} + w_3 a_{31} \\
L_2 &= w_0 + w_1 a_{12} + w_2 a_{22} + w_3 a_{32} \\
&\vdots \\
L_N &= w_0 + w_1 a_{1N} + w_2 a_{2N} + w_3 a_{3N}
\end{aligned}
$$

where a_{ij} is the jth sample of the ith attribute. Notice that the above equations can be written in matrix form as follows.

$$
\begin{bmatrix} L_1 \\ L_2 \\ L_3 \\ L_4 \end{bmatrix} = \begin{bmatrix} 1 & a_{11} & a_{21} & a_{31} \\ 1 & a_{12} & a_{22} & a_{32} \\ \vdots & \vdots & \vdots & \vdots \\ 1 & a_{1N} & a_{2N} & a_{33} \end{bmatrix} \begin{bmatrix} w_0 \\ w_1 \\ w_2 \\ w_3 \end{bmatrix} \tag{7.22}
$$

Equation 7.22 can be expressed in the notation form as follows.

$$
L = AW \tag{7.23}
$$

where L is an $N \times 1$ matrix containing the known log values, A is a $N \times 4$ matrix containing the attribute values, and W is a 4×1 matrix with the unknown weights (Taner 1994). This can be solved by least-squares minimization in the following way.

$$W = \left[A^T A\right]^{-1} A^T L \tag{7.24}$$

As a detailed computation, the weights can be estimated in the following way.

$$\begin{bmatrix} w_0 \\ w_1 \\ w_2 \\ w_3 \end{bmatrix} = \begin{bmatrix} N & \sum a_{1i} & \sum a_{2i} & \sum a_{3i} \\ \sum a_{1i} & \sum a_{1i}^2 & \sum a_{1i}a_{2i} & \sum a_{1i}a_{3i} \\ \sum a_{2i} & \sum a_{1i}a_{2i} & \sum a_{2i}^2 & \sum a_{2i}a_{3i} \\ \sum a_{3i} & \sum a_{1i}a_{3i} & \sum a_{2i}a_{3i} & \sum a_{3i}^2 \end{bmatrix}^{-1} \begin{bmatrix} \sum L_i \\ \sum a_{1i}L_i \\ \sum a_{2i}L_i \\ \sum a_{3i}L_i \end{bmatrix} \tag{7.25}$$

Just as in the single-attribute case, the mean-squared error (7.23) calculated using the derived weights constitutes a goodness-of-fit measure for the transform, as does the normalized correlation defined in Eq. (7.14), where the x-coordinate is now the predicted log value and the y-coordinate is the real log value (Russell et al. 1997; Swisi 2009).

7.4.1 Determining Attributes by Stepwise Regression

We derived equations in the previous sections that enable us to identify optimal operators for any set of attributes. These operators are optimal in the sense of minimizing the mean-squared error of prediction between the actual target logs and the predicted target logs. The next issue to be dealt with is how these attributes are selected.

There are two approaches available to select attributes. The first approach is to exhaustive search and the second approach is stepwise regression methods. The first approach is very much time-consuming and hence less used. The second approach i.e. stepwise regression although less optimal, a procedure developed by Draper and Smith in (1966). In this procedure, the assumption is that if you already know the best combination of M attributes, then the best combination of $M + 1$ attributes includes the previous M attributes as members. Of course, it is necessary to rederive the previously calculated coefficients. The process is shown in the section below.

First, by exhaustive search, find the single best attribute and solve the optimum coefficients and calculate the error of prediction. The best attribute is the one with the lowest prediction error. Let say this attribute to be a_1.

Second, find the best attribute pair, assuming that attribute 1 (a_1) is the first member. Form all pairs for each other attribute in the list, solve the optimum coefficients and calculate the error of prediction. The best pair is the one with the lowest error of prediction. Let say the second attribute of the best pair is a_2.

Similarly, in the third step, find the best triplet of attributes, assuming that the first two members are a_1 and a_2. For each other attribute in the list, all triplets are formed and solve the optimum coefficients for each triplet and calculate the error of prediction. The best triplet is a_3.

Continue as long as you desired this process. The first thing to note is that this process's calculation time is much shorter than exhaustive searching (Russell et al. 1997, 2001). The problem with stepwise regression is that one cannot be sure that the optimal solution will be derived. In other words, the combination of five attributes found may not be the best five by exhaustive search. Each additional attribute found, however, has an error of prediction less than or equal to the combination found previously. If the new prediction error is greater than the combination of the previous set, all the weights for this new attribute to be zero and the prediction error is equal to the previous set.

7.4.2 Application of MAR

An example of the application of multi-attribute regression to predict porosity is presented for Blackfoot data from Alberta, Canada to get a real feel about how the method work. The list of attributes is generated from seismic data directly or indirectly with a 7-point convolution operator and some of them are presented in Table 7.2. Each line of the table contains information for a particular attribute and they are arranged in order of increasing prediction error. In this table, the training error corresponds to a combination of attributes with the ones above it. On the other hand, the validation column of the table represents the cross-validation error. As discussed earlier, in standard condition, the training error should decrease as one add new attributes (Pramanik et al. 2004; Eskandari et al. 2004) but one can notice that by adding the fourth attribute i.e. *ApparentPolarity*, the validation error increases and hence all the weights corresponding to this attributes become to zero. The graphical representation of Table 7.2 is given in Fig. 7.12. This analysis restricts to selected starting three attributes as the desired and they are 1/(*Impedance*), *Amplitude Weighted Cosine Phase* and *X-Coordinate*.

Table 7.2 Multi-attribute list and their error with target porosity

S. No.	Target	Final attribute	Training error	Validation error
1.	Porosity	1/(Impedance)	4.963269	5.046682
2.	Porosity	Amplitude weighted cosine phase	4.813295	4.959127
3.	Porosity	X-Coordinate	4.77781	4.925902
4.	Porosity	Apparent polarity	4.745512	4.95469
5.	Porosity	Instantaneous frequency	4.710768	4.973616
6.	Porosity	Dominant frequency	4.684892	4.982152
7.	Porosity	Filter 25/30–35/40	4.659465	4.991252

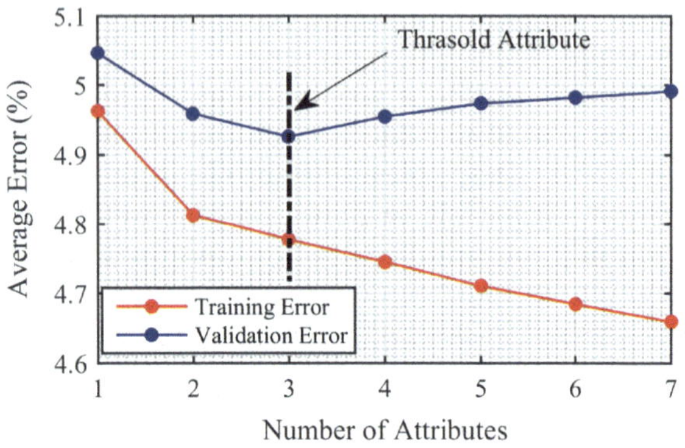

Fig. 7.12 Prediction error plot using all attributes for multi-attribute regression

The desired attributes are then cross plotted against the well log porosity and a best fit straight line provides the relation between attributes and porosity at a particular sample point which is further utilized to predict porosity in the inter-well region. The process is similar to the single attribute analysis but the only difference is that in multi-attribute regression uses more than one attribute at a time. The MAR is first applied to the composite trace near to well locations and porosity is predicted. Although the porosity is already known at a well location to cross-verify predicted porosity, the composite trace analysis needs to be performed. Figure 7.13 shows a comparison of predicted porosity (red lines) with actual porosity (black lines) for well 01-17, 08-08 and 16-08. From the figure, one can notice that the predicted porosity is followed the trend to the original porosity for all wells. The average correlation between the two is estimated to be 0.58 and error 4.61. Further, a cross plot is generated between actual porosity from well log data and predicted porosity and displayed in Fig. 7.14. The distribution of scatter points shows that the predicted porosity closes to the original one. The composite trace analysis shows quite good prediction results and hence one can go for 3D porosity prediction.

Now that we have derived the relationship between attributes and target logs, one can apply the analysis to the whole 3D volume. A cross-section of predicted 3D porosity volume at inline 27 is presented in Fig. 7.15. The predicted section shows very high-resolution subsurface images as compared to the input seismic section. The well log porosity from well 08-08 is also plotted on the inverted section and shows a good match between them. The close inspection of the inverted section shows an anomaly zone with very high porosity ranging from 0.12 to 0.15 particularly in the time interval between 1055 ms and 1070 ms two way travel time. This high porosity is correlated with low impedance anomaly interpreted in chapter 4 and hence interpreted as Glauconitic sand channel. A 3D volume of predicted porosity is displayed in Fig. 7.16. The figure highlighted the anomalous zone in all directions in the subsurface.

The benefit of the 3D section is that one can analyze the anomalous zone vertically as well as horizontally. Till now, the discussion is performed for vertical variation of the sand channel but now, one can generate a horizontal variation of the channel to

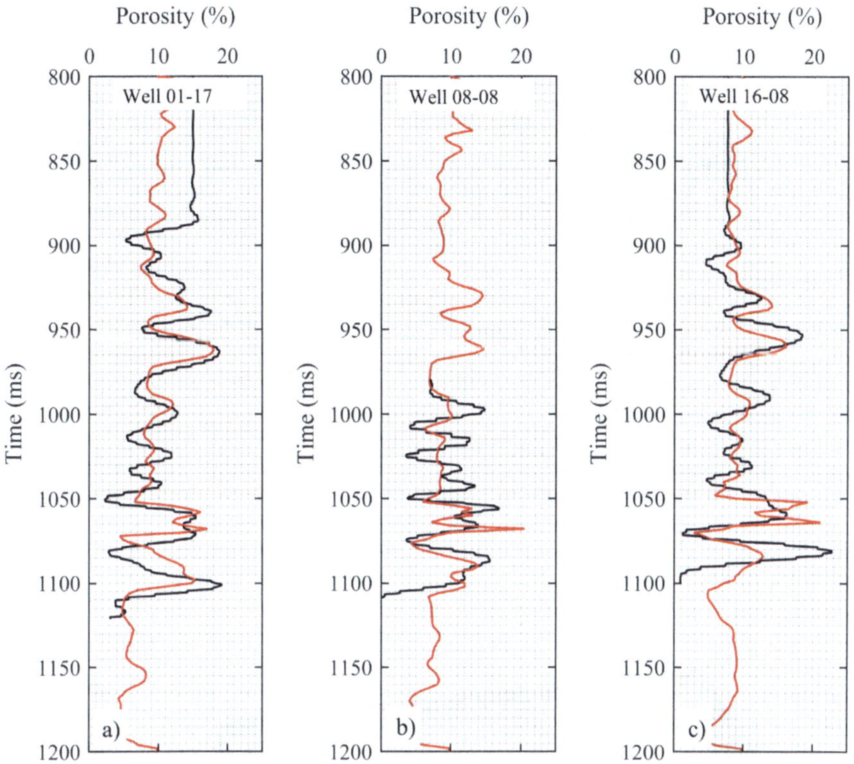

Fig. 7.13 Comparison of predicted porosity with the actual porosity from the **a** well 01-17, **b** well 08-08, and **c** well 16-08

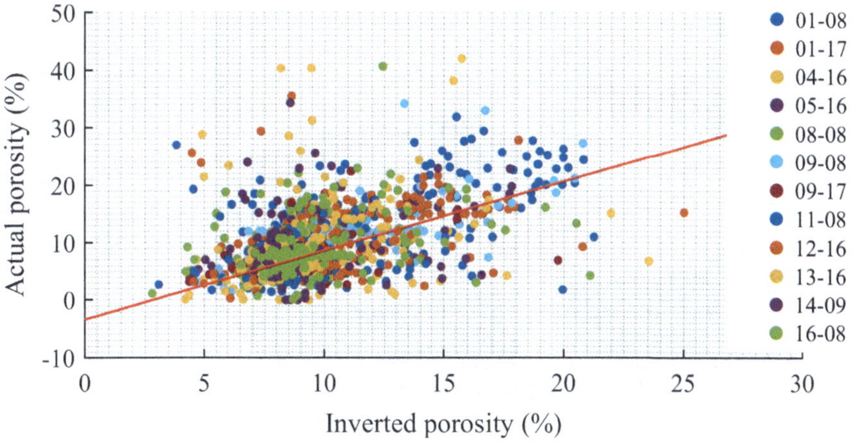

Fig. 7.14 Crossplot of predicted porosity versus actual porosity for all the 13 wells. The red solid line indicates the best fit line

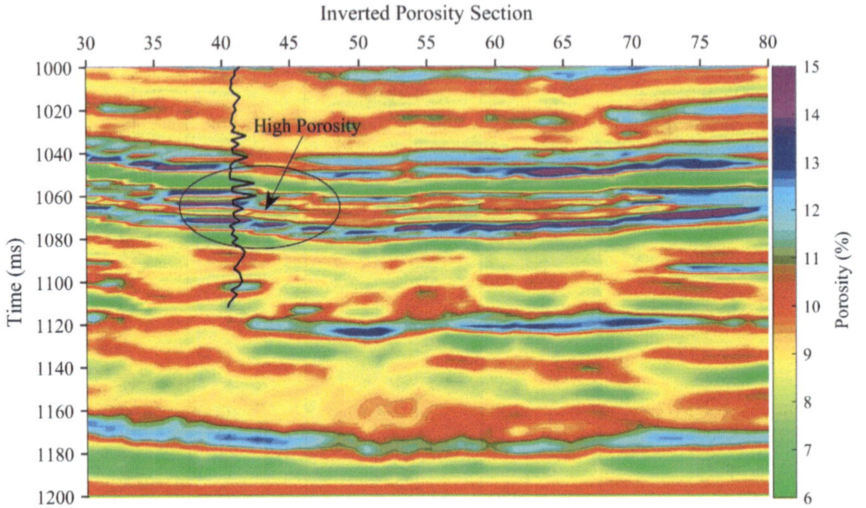

Fig. 7.15 Cross-section of predicted porosity at inline 61. The anomalous zone (high porosity) is highlighted by the ellipse

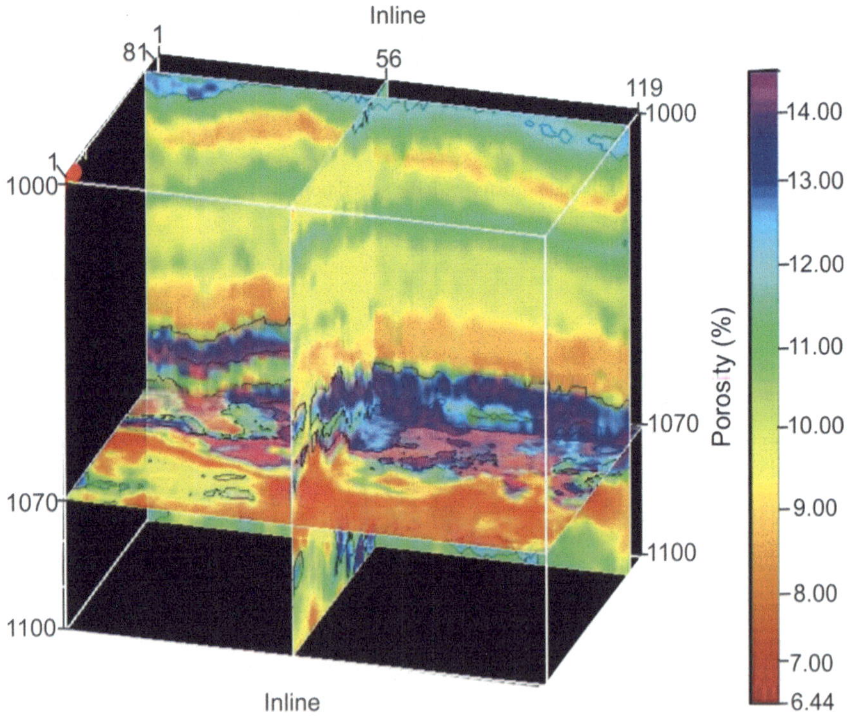

Fig. 7.16 3D representation of predicted porosity of the Blackfoot field, Canada

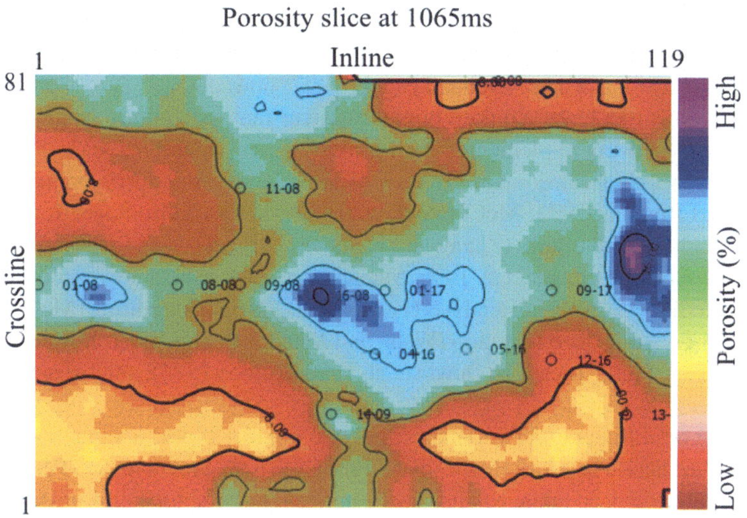

Fig. 7.17 Slice at 1065 ms two way travel time of predicted porosity

monitor subsurface features. A slice at 1065 ms time intervals is shown in Fig. 7.17 which shows the variation of porosity in the subsurface. One can notice that the horizontal variation of porosity is nicely visible and can be used to analyze the details of the sand channel. This slice indicates the variation of the sand channel by blue to pink hot color. From this analysis, the width of the anomalous zone horizontally can also be estimated. It is also noticed that the sand channel varies from north-east (NE) to south-west (SW) direction. As the seismic data shows amplitude variation along with time and cannot be interpreted subsurface lithology whereas by using the above techniques one can obtain the distribution of petrophysical parameters in the inter-well region. In this example, porosity prediction is performed but the other petrophysical properties like permeability, saturation, shale sand ratio, gamma-ray, etc. can also be generated which helps to decide movable hydrocarbon saturation, net pay, etc. This technique is very helpful for the exploration and production projects.

7.5 Neural Network Techniques

The assessment was linear so far. A linear straight line was fitted to the cross plot between the target logs and attribute(s) and weights are calculated for the regression line by minimizing the mean-square error. But, you might think a greater order curve would fit better with the points rather than fitting a straight line. There are several choices to calculate this curve. One choice is to apply a nonlinear transformation to either or both of the variables and to fit a straight line to the transformed data (Schultz et al. 1994). A second choice is to match a polynomial of greater order. A third choice is also available which is to use a neural network to obtain the relationship between target logs and attributes. In geophysics, neural networks are becoming increasingly common (McCormack 1991; Schuelke et al. 1997) nowadays.

Neural networks have acquired popularity over the past century in geophysics. They were effectively implemented to a verity of issues. Neural networks have been used in the geophysical domain for waveform recognition and first-break pick-up (Murat and Rudman 1992; McCormack et al. 1993); for electromagnetic (Poulton et al. 1992), magnetotelluric (Zhang and Paulson 1997) and seismic inversion (Röth and Tarantola 1994; Langer et al. 1996; Macías et al. 1998); shear-wave splitting (Dai and MacBeth 1994), well-log analysis (Huang et al. 1996), trace editing (McCormack et al. 1993), seismic deconvolution (Wang and Mendel 1992; Calderón-Macías et al. 1997), and event classification (Dowla et al. 1990; Romeo 1994); and for many other problems.

Many types of neural networks exist but two popular types' namely multi-layer feed-forward neural network (MLFN) and Probabilistic neural network (PNN). Both methods have their advantages and disadvantages to the application of data. Both methods are described briefly in the following sections along with real data examples.

7.5.1 Multi-layer Feed Forward Neural Network (MLFN)

In this section, we limit our discussion to the static feed-forward neural networks (MLFN). It is called a static algorithm because the weights estimated during the process are fixed and do not change with time. The feed-forward indicates that the output is not feedback, i.e., refed, to the network. Thus, this type of network does not iterate to a final solution but directly translates the input signals to output independent of previous input.

McCulloch and Pitts (1943) conceived the mathematical perceptron some 55 years ago to mimic the behavior of the biological neuron (Fig. 7.18). The biological neuron consists primarily of three components/layers, one input layer, one or more hidden layers, and one output layer. Each layer of the network consists of nodes and connects these nodes with each other. These connections represent weights. These weights are the final element of the production decision as they play a very significant part in the implementation of MLFN. Figure 7.18 shows the MLFN's network architecture. The amount of input nodes is determined by the amount of characteristics during the MLFN phase, and one chooses to use them to forecast rock characteristics (Maurya and Singh 2019). The number of attributes is determined by the convolution operator process in which a specific combination of attributes has been tested and the combination with maximum similarity is selected against the desired log property (Russell et al. 2001; Svozil et al. 1997).

MLF neural networks are the most common neural networks, trained with a learning algorithm for back-propagation. For the formal description of the neurons, we can use the so-called mapping function Γ, that assigns for each neuron i a subset $\Gamma(i) \subseteq V$ which consists of all ancestors of the given neuron. A subset $\Gamma^{-1}(i) \subseteq V$

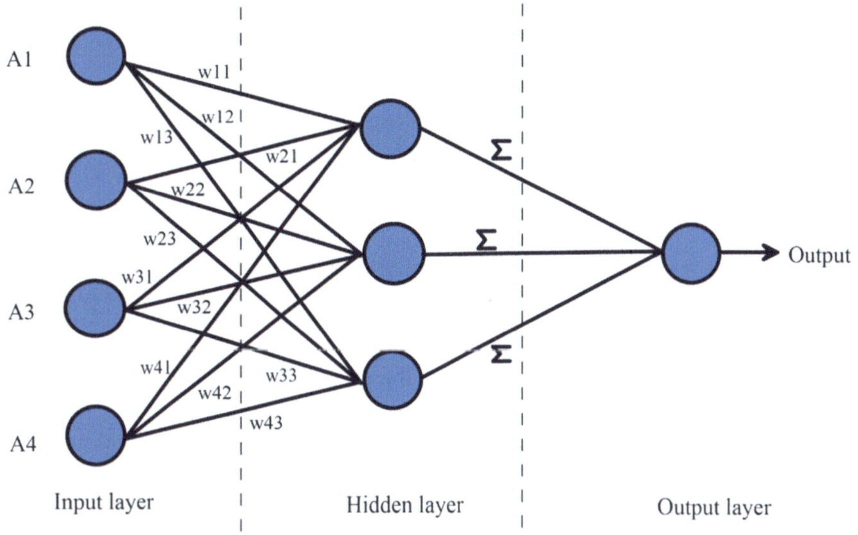

Fig. 7.18 Representation of MLFN architecture. Ai indicates desired attributes, wij indicated weights

than consists of all predecessors of the given neuron i. Each neuron in a particular layer is connected with all neurons in the next layer. The connection between the ith and jth neuron is characterized by the weight coefficient ω_{ij} and the ith neuron by the threshold coefficient v_i. The weight coefficient reflects the degree of importance of the given connection in the neural network (Svozil et al. 1997). The output value (activity) of the ith neuron x_i is determined by Equations given below.

$$x_i = f(\psi_i) \tag{7.26}$$

$$\psi_i = v_i + \sum_{j \in \Gamma_i^{-1}} \omega_{ij} x_j \tag{7.27}$$

where ψ_i is the potential of the ith neuron and function $f(\psi_i)$ is the so-called transfer function. The threshold coefficient can be understood as a weight coefficient of the connection with formally added neuron j, where $x_j = 1$.

The transfer function can be written as follows.

$$f(\psi_i) = \frac{1}{1 + \exp(-\psi)} \tag{7.28}$$

The supervised adaptation process varies the threshold coefficients v_i and weight coefficients ω_{ij} to minimize the sum of the squared differences between the computed and required output values (Luo et al. 2000). This is accomplished by minimization of the objective function:

$$E = \sum_{p} \frac{1}{2} \left(x_p - \hat{x}_p \right)^2 \tag{7.29}$$

where x_p and $\hat{x}_p$ are vectors composed of the computed and required activities of the output neurons and summation runs over all output neurons p.

7.5.2 Training and Generalization of MLFN

The neural network works in two modes, first is training and the second is the prediction. You need two information sets, the practice set and the set you want to predict (test set) to train the neural network and to predict.

The training mode starts with arbitrary weight values that could be random numbers and iteratively proceeds. An epoch is called every iteration of the full training set. The network adjusts the weights in the direction that decreases the error in each epoch. The weights gradually converge to the locally optimal set of values as the iterative method of incremental adjustment proceeds. Before training is finished, many epochs are generally needed (Svozil et al. 1997; Wu et al. 1992).

Back-propagation learning can be carried out in one of two fundamental ways for a specified training set i.e. pattern mode and batch mode. Weight updating is conducted after each training pattern is presented in the pattern mode of teaching backpropagation. Weight updating is conducted in the back-propagation learning batch mode after all the training examples are presented (i.e. after the whole epoch). The pattern mode is preferred over the batch mode from an 'on-line' point of perspective, as it needs less local storage for each synaptic connection. In addition, given that patterns are randomly presented to the network, the use of patterns by pattern updating weights makes the search in weight space stochastic, making it less likely that the backpropagation algorithm will be trapped in a local minimum. On the other side, the use of training batch mode offers a more precise gradient vector estimate. For example, pattern mode should be used in online process control, as not all training patterns are available in the given time. Ultimately, the comparative efficacy of the two training methods relies on the issue solved (Russell et al. 1997).

Information flows through the network in forecast mode, from inputs to outputs. The network processes one example at a time, producing an estimate based on the input values of the output value(s). The resulting error is used as an estimate of the trained network's predictive quality. One can begin with a practice set in back-propagation learning and use the back-propagation algorithm to calculate the network's synaptic weights (Tonn 2002; Iturrarán and Parra 2014). The hope is that the so-called neural network will become widespread. A network is said to generalize well when the network-computed input-output relationship is right (or almost right) for input/output models that have never been used in-network practice. Generalization is not a mystical property of neural networks but can be contrasted with the impact of excellent non-linear input information interpolation. When the learning method repeats too many iterations (i.e. the neural network is over-trained or over-fitted, there is no distinction between overtraining and overfitting), the network can memorize the training data and thus be less willing to generalize comparable patterns of input-output (Leite and Vidal 2011; Mahmood et al. 2017). The network provides almost ideal outcomes from the practice set for examples but fails from the test set for example. Overfitting can be contrasted with unsuitable polynomial degree selection in polynomial regression. With noisy data, severe overfitting can happen even when there are many more instances of training than weights. A sufficiently big set of training instances is the fundamental condition for excellent generalization. This training set must be representative of the set of all cases to which you wish to generalize at the same time. The significance of this situation is due to two kinds of generalization: interpolation and extrapolation (Masters 1995; Masri et al. 2000; Singh et al. 2016). Interpolation applies to instances encircled more or less by neighboring instances of practice; all else is extrapolation. In specific, instances outside the training information spectrum involve extrapolation. Interpolation can often be performed reliably, but it is notoriously unreliable to extrapolate. Therefore, adequate training information is essential to prevent the need for extrapolation. Experimental design results in methods for choosing excellent training sets. Because of a set quantity of training data, there are some efficient methods to avoid overfitting and generalization (Wu et al. 1992).

7.5.3 Application of MLFN

An example from the Blackfoot field, Canada is presented here to increase the understanding of MLFN methods. As in the earlier case, the MLFN also applied to the data, first is for composite trace and the second is for entire 3D datasets. For this purpose, one composite trace extracted from near to well 08-08 and thereafter the MLFN is applied on it to predict porosity trace. Figure 7.19 shows a comparison

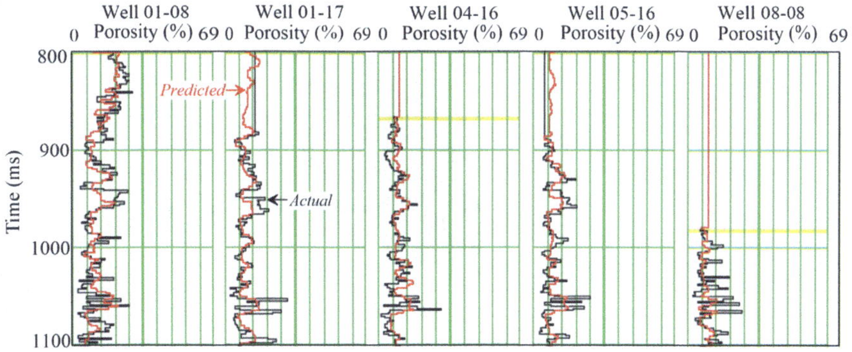

Fig. 7.19 Comparison of well log porosity with actual porosity estimated using MLFN methods

of predicted porosity and actual porosity. It is noticed from the Fig. 7.19 that the predicted porosity is following the trend with actual porosity quite nicely. The average correlation coefficient is estimated to be 0.59 and RMS error is to be 4.7. These values show that the MLFN technique is quite good enough to estimate subsurface rock property although the correlation is low and error is high as compared with the MLFN methods. This is not always true because the MLFN largely depends on input parameters. If one can able to choose the perfect combination of input parameters, then the methods can show better results as compared with multi-attribute regression. In this example, the analysis is presented to understand how MLFN works, not to compare with other geostatistical methods discussed here. The analyses were performed in different conditions and considering different input parameters hence cannot compare explicitly with each other.

A cross plot between predicted porosity and actual porosity is also generated and shown in Fig. 7.20. The distribution of scatter points demonstrates that the predicted porosity is close to the actual porosity.

To analysis the 3D variation of porosity, the MLFN can be performed for the entire seismic volume. It is time taking but can provide a detailed picture of the subsurface. A cross-section at inline 27 is presented in Fig. 7.21. The inverted section shows very high-resolution subsurface layers. Well, log porosity from well 08-08 is also plotted on the predicted section. One can notice that the predicted porosity and well log porosity matching very nicely with each other. The layer of the subsurface can be easily identified with this section. The interpretation of the predicted porosity section shows many patches of high porosity ranging from 0.12 to 0.15. One patch of high porosity is matching with a well log anomaly zone which falls in between 1055 and 1070 ms time interval and predicted as a sand channel. This high porosity anomaly

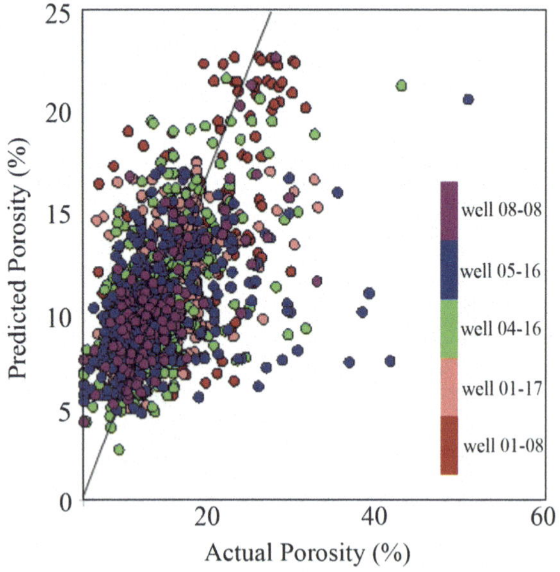

Fig. 7.20 Crossplot of well log porosity (well) versus modeled porosity. The color bar shows two way travel time

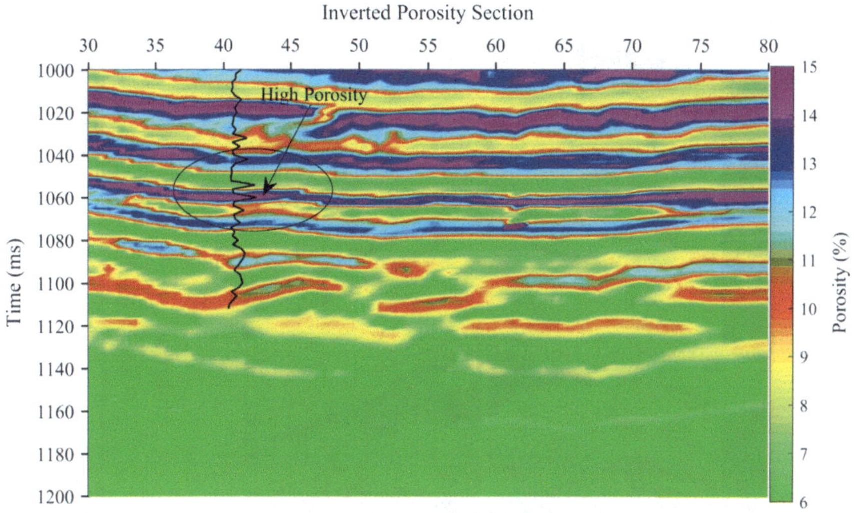

Fig. 7.21 Cross-section of predicted porosity at inline 27 estimated using MLFN methods. The anomalous zone is highlighted by the ellipse

is corroborating well with the low impedance anomaly and high porosity anomaly estimated with other methods discussed so far. A 3D plot of predicted porosity is provided in Fig. 7.22 to get the variation of the sand channel vertically as well as horizontally. Further, to measure the horizontal variation of the sand channel, slice at 1065 ms time intervals are generated from predicted porosity volume and shown in Fig. 7.23. The variation of the sand channel is highlighted by the arrow. It is noticed from the figure that the sand channel varies from NE to SW direction which is already shown for multi-attribute analysis cases.

7.5.4 Probabilistic Neural Network (PNN)

A probabilistic neural network is an alternative form of neural network. In fact, the probabilistic neural network (PNN) is a mathematical interpolation system that uses neural network architecture to implement it. This is a potential benefit, as you can often comprehend their conduct much better than the MLFN by learning the mathematical formulation (Specht 1990, 1991; Masters 1995).

The fundamental concept behind the general probabilistic neural network (Specht 1991; Masters 1995) is to use one or more measured values called independent variables to predict the value called the dependent variable. Let's an independent variable is represented by vector $x = [x_1, x_2, \ldots, x_p]$, where p is the number of independent variables. On the other hand, the dependent variable L is a scalar. The PNN takes variables $x_1, x_2, \ldots, x_p$ as input and generate y as output. The objective

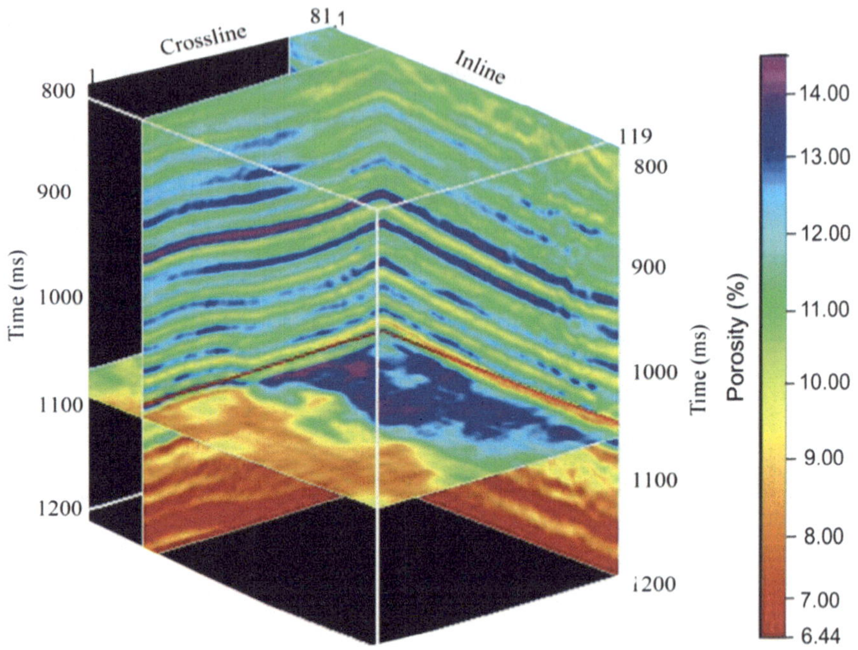

Fig. 7.22 3D view of predicted porosity estimated using MLFN methods

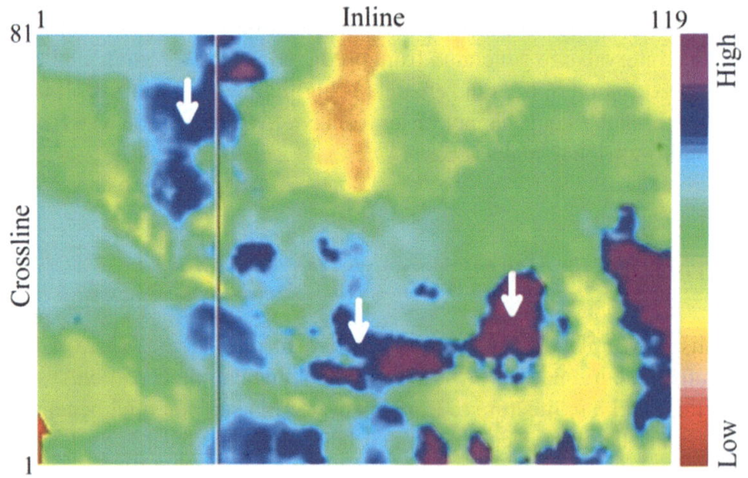

Fig. 7.23 Slice at channel interval (1065 ms) highlighted channel trend

is to predict the unknown dependent variable, $\hat{L}$ at a place where the independent variables are known. This estimate is based on the basic equation of the probabilistic neural network of the general regression given below.

$$\hat{L}(x) = \frac{\sum_{i=1}^{n} L_i \exp(-D(x, x_i))}{\sum_{i=1}^{n} \exp(-D(x, x_i))} \tag{7.30}$$

In Eq. 7.30, n represents the number of examples and $D(x, x_i)$ is the distance between the input point and each of the training points x_i. The quantity $D(x, x_i)$ can be calculated as follows.

$$D(x, x_i) = \sum_{j=1}^{3} \left(\frac{x_j - x_{ij}}{\sigma_j} \right)^2 \tag{7.31}$$

The $D(x, x_i)$ is measured in the multidimensional space spanned by the attributes and is scaled by the quantity σ_j, which may differ for each of the attributes (Specht 1990).

Equations (7.30) and (7.31) describe the application of the PNN network whereas the training of the network consists of determining the optimal set of smoothing parameters called σ_j. The criterion for determining these parameters is to have the smallest validation error in the resulting network (Leiphart and Hart 2001). The total validation error for the *m*th the target sample is defined as follows.

$$E_V(\sigma_1, \sigma_2, \sigma_3) = \sum_{i=1}^{N} \left(L_i - \hat{L}_i \right)^2 \tag{7.32}$$

This can be also noted that the prediction error depends largely on the choice of the parameters σ_j. The validation error is minimized using a nonlinear conjugate gradient algorithm with respect to the smoothing parameters.

7.5.5 Application of PNN

An example of PNN is presented here to predict porosity for Blackfoot datasets, Canada. The process of PNN is nearly the same as for multi-attribute analysis cases with the only difference is that a nonlinear relationship is set up between attributes and well log porosity rather linear relationship. The list of attributes generated for the multi-attribute case (Table 7.2) is used for the PNN case also. After selecting desired attributes, the nonlinear relationship is set up at each sample points which is used to predict porosity at inter-well region. First, the prediction is performed for the composite trace and the result is shown in Fig. 7.24. The results are shown for well 01-17, 08-08 and well 16-08. In the analysis of PNN, the impedance derived from maximum likelihood inversion was used as one attribute. From the figure, it

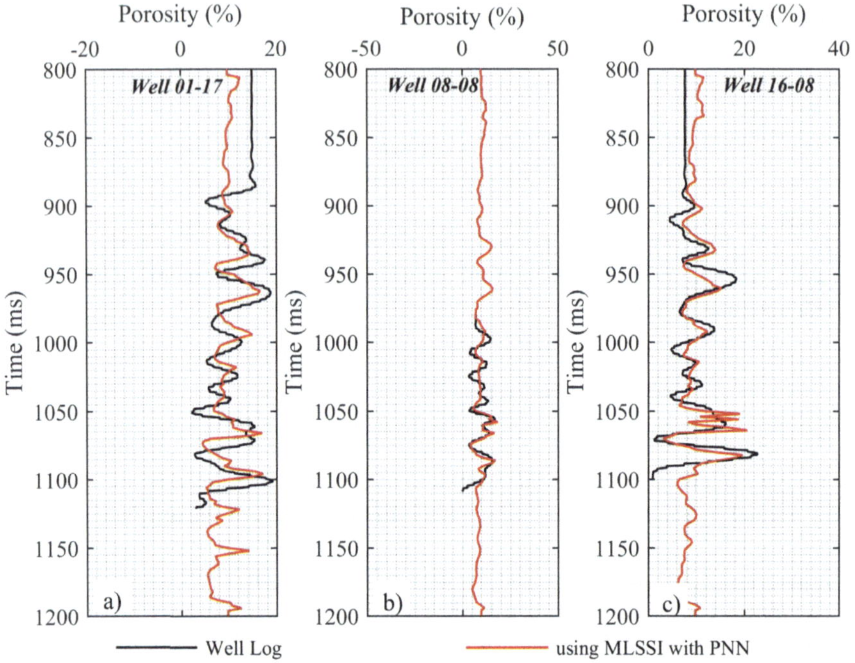

Fig. 7.24 Comparison of predicted porosity with actual porosity estimated using PNN methods

is noticed that the predicted curves are following original curves very nicely for all wells. Further, the cross-plot between predicted porosity and original porosity is generated and shown in Fig. 7.25. The distribution of scattered points shows that the inverted porosity is very close to the actual porosity which is also evident by the correlation coefficient. The PNN shows superior results, i.e. lower prediction error and lower validation error. The correlation coefficient is estimated to be 0.76 and RMS error is 3.93, indicating a good correlation and RMS. One can argue that the probabilistic neural network predicts the logs with higher accuracy. The multi-regression analysis predicted the logs with correlation 0.59 while the PNN predicted them with correlation 0.76. A comparison of prediction error with validation error for all wells are presented in Fig. 7.26. Both errors vary in a similar way with an average value of 3.93.

Once the relation between the seismic attributes and the porosity records is determined, the PNN is applied to the data volumes. Figures 7.27 demonstrate a porosity cube at inline 27. The well log porosity is also plotted over the predicted porosity and it is found that both the porosities are matching very well. It is very possible to distinguish the sand channel as a large anomaly of porosity. Note again the greater resolution obtained through the use of the probabilistic neural network. The 3D view of predicted porosity is shown in Fig. 7.28 and three-dimensional variation of the sand channel are can be noticed with high porosity contrast. Figure 7.29 shows porosity slice at the sand channel level and shows the horizontal distribution of the sand

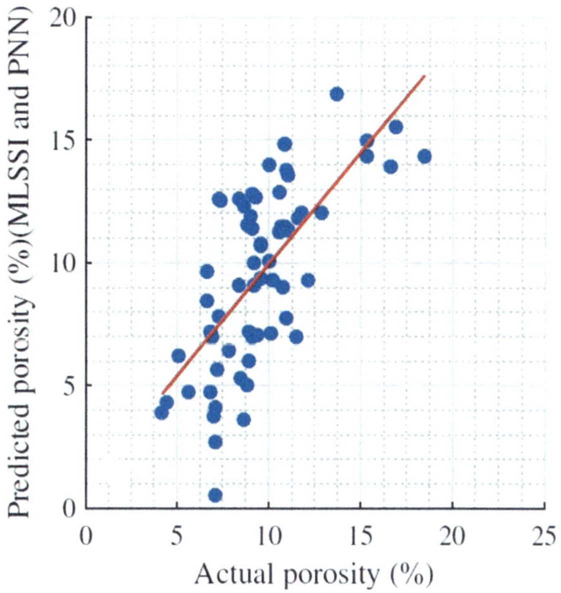

Fig. 7.25 Cross plot of actual porosity and predicted porosity for well 08-07

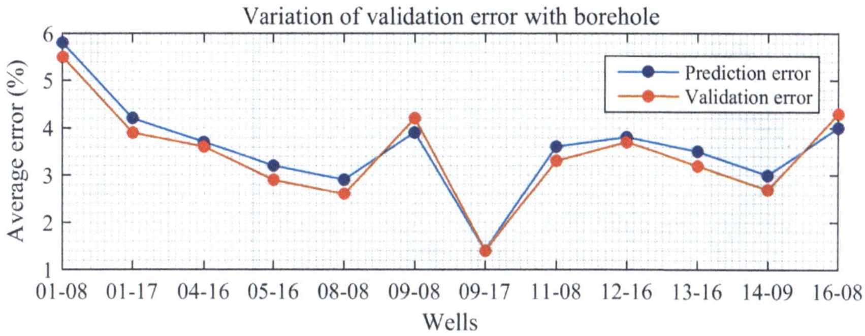

Fig. 7.26 Variation of Prediction and validation error with well logs

channel. The high porosity zone in the seismic section is shown in Fig. 7.30. This figure nicely highlighted the anomalous zone. From this analysis, one can easily interpret the seismic section and can identify the visualization of the sand channel. Apart from that, these sections can also be used to derive a relationship between them. For example, the inverted impedance and predicted porosity section can be cross plotted and can be used to estimate the relation between them by fitting a straight line or higher-order polynomial. This relationship will be more meaningful as compared with the relation derived from the well log curves at well location. The well log derived equation provides a relationship at a particular location that cannot

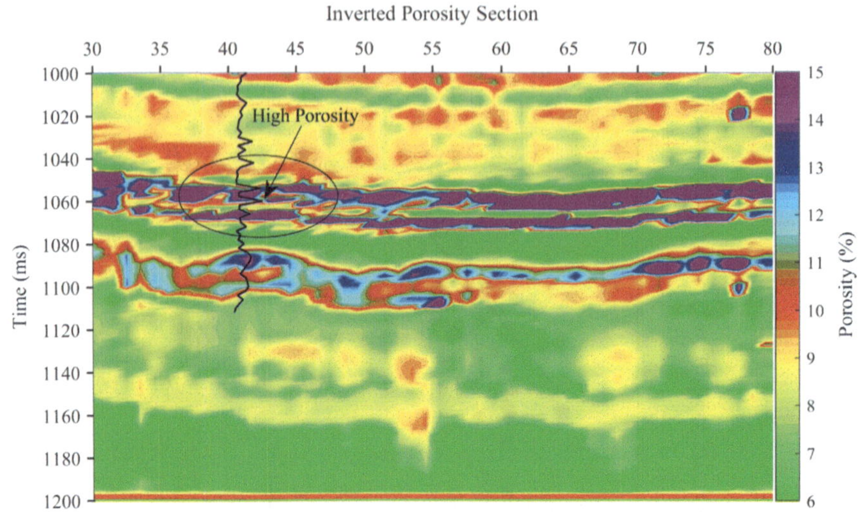

Fig. 7.27 Cross-section of predicted porosity at inline 27. The anomalous zone is highlighted by an ellipse

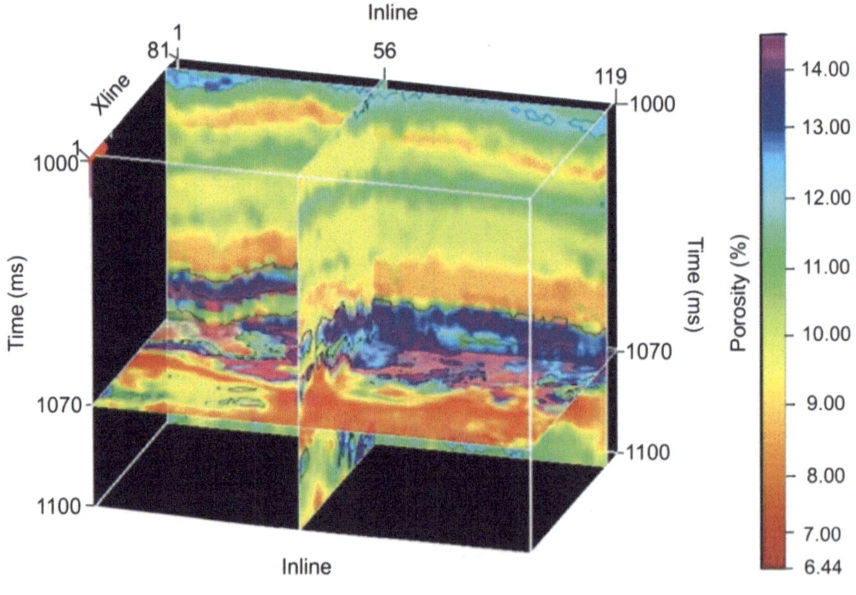

Fig. 7.28 3D view of predicted porosity. The anomalous zone is shown in pink color

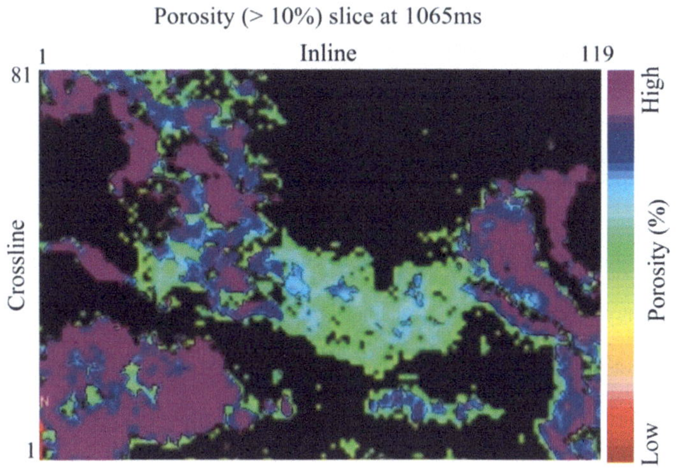

Fig. 7.29 Variation of porosity (>10%) in the subsurface at 1065 ms TWT

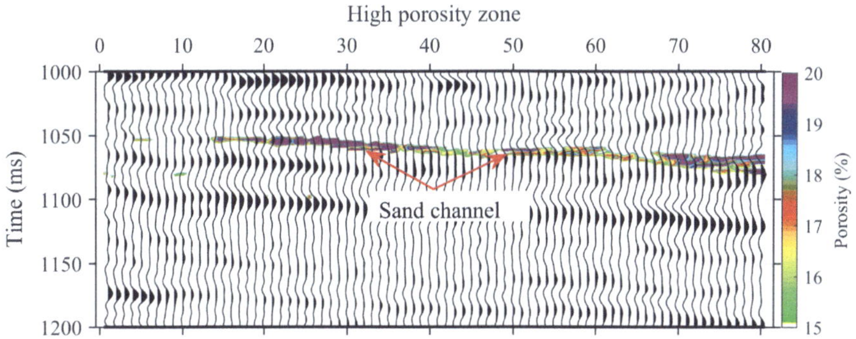

Fig. 7.30 Seismic cross-section highlighted the sand channel. The color bar indicates porosity

valid away from the well whereas the relationship derived from the cross-section is valid anywhere in the region and give a better estimate (Maurya et al. 2017) of the subsurface property.

References

Adeli H, Panakkat A (2009) A probabilistic neural network for earthquake magnitude prediction. Neural Netw 22(7):1018–1024

Anderson JK (1996) Limitations of seismic inversion for porosity and pore fluid: lessons from chalk reservoir characterization and exploration. In: SEG technical program expanded abstracts, society of exploration geophysicists, pp 309–312

Barnes AE (1991) Instantaneous frequency and amplitude at the envelope peak of a constant-phase wavelet. Geophysics 56(7):1058

Barnes AE (1993) Instantaneous spectral bandwidth and dominant frequency with applications to seismic reflection data. Geophysics 58(3):419–428
Barnes AE (1994) Theory of two-dimensional complex seismic trace analysis. In: SEG technical program expanded abstracts, society of exploration geophysicists, pp 1580–1583
Bosch M, Mukerji T, Gonzalez EF (2010) Seismic inversion for reservoir properties combining statistical rock physics and geostatistics: a review. Geophysics 75(5):75A165–75A176
Brown AR (1996) Seismic attributes and their classification. Lead Edge 15(10):1090
Calderón-Macías C, Sen MK, Stoffa PL (1997) Hopfield neural networks, and mean-field annealing for seismic deconvolution and multiple attenuations. Geophysics 62(3):992–1002
Chambers RL, Yarus JM (2002) Quantitative use of seismic attributes for reservoir characterization. CSEG Recorder 27(6):14–25
Chiles J (1988) Fractal and geostatistical methods for modeling of a fracture network. Math Geol 20(6):631–654
Chopra S, Marfurt KJ (2005) Seismic attributes—a historical perspective. Geophysics 70(5):3SO–28SO
Chopra S, Marfurt KJ (2007) Seismic attributes for prospect identification and reservoir characterization. Society of Exploration Geophysicists and European Association of Geoscientists and Engineers
Cohen L, Lee C (1990) Instantaneous bandwidth for signals and spectrograms. In: International conference on acoustics, speech, and signal processing, IEEE, pp 2451–2454
Dai H, MacBeth C (1994) Split shear-wave analysis using an artificial neural network. The first Break 12(12):605–613
Dowla FU, Taylor SR, Anderson RW (1990) Seismic discrimination with artificial neural networks: preliminary results with regional spectral data. Bull Seismol Soc Am 80(5):1346–1373
Doyen PM (1988) Porosity from seismic data: a geostatistical approach. Geophysics 53(10):1263–1275
Draper N, Smith H (1966) Applied regression analysis. Wiley, New York
Dubrule O (2003) Geostatistics for seismic data integration in Earth models. Distinguished instructor short course. Number 6. SEG Books
Eskandari H, Rezaee MR, Mohammadnia M (2004) Application of multiple regression and artificial neural network techniques to predict shear wave velocity from well log data for a carbonate reservoir, south-west Iran. CSEG Recorder 29(7):42–48
Haas A, Dubrule O (1994) Geostatistical inversion- a sequential method of stochastic reservoir modeling constrained by seismic data. First Break 12(11):561–569
Hampson D, Todorov T, Russell B (2000) Using multi-attribute transforms to predict log properties from seismic data. Explor Geophys 31(3):481–487
Hampson DP, Schuelke JS, Quirein JA (2001) Use of multiattribute transforms to predict log properties from seismic data. Geophysics 66(1):220–236
Huang Z, Shimeld J, Williamson M, Katsube J (1996) Permeability prediction with artificial neural network modeling in the Venture gas field, offshore eastern Canada. Geophysics 61(2):422–436
Iturrarán VU, Parra JO (2014) Artificial neural networks applied to estimate permeability, porosity and intrinsic attenuation using seismic attributes and well-log data. J Appl Geophys 107:45–54
Jones TA, Helwick SJ (1998) Method of generating 3-d geologic models incorporating geologic and geophysical constraints. US Patent 5:634–838
Langer H, Nunnari G, Occhipinti L (1996) Estimation of seismic waveform governing parameters with neural networks. J Geophys Res Solid Earth 101(B9):20109–20118
Leiphart DJ, Hart BS (2001) Comparison of linear regression and a probabilistic neural network to predict porosity from 3-D seismic attributes in Lower Brushy Canyon channeled sandstones, southeast New Mexico. Geophysics 66(5):1349–1358
Leite EP, Vidal AC (2011) 3D porosity prediction from seismic inversion and neural networks. Comput Geosci 37(8):1174–1180
Lindseth RO (1979) Synthetic sonic logs a process for stratigraphic interpretation. Geophysics 44(1):3–26

Luo X, Patton AD, Singh C (2000) Real power transfer capability calculations using multi-layer feed-forward neural networks. IEEE Trans Power Syst 15(2):903–908

Macías D, Pérez-Pomares JM, García-Garrido L, Carmona R, Muñoz-Chápuli R (1998) Immunoreactivity of the ETS-1 transcription factor correlates with areas of epithelial-mesenchymal transition in the developing avian heart. Anat Embryol 198(4):307–315

Mahmood MF, Shakir U, Abuzar MK, Khan MA, Khattak N, Hussain HS, Tahir AR (2017) A probabilistic neural network approach for porosity prediction in the Balkassar area: a case study. Journal of Himalayan Earth Science 50:1

Masri S, Smyth A, Chassiakos A, Caughey T, Hunter N (2000) Application of neural networks for the detection of changes in nonlinear systems. J Eng Mech 126(7):666–676

Masters T (1995) Advanced algorithms for neural networks: a C++ sourcebook. Wiley, Hoboken

Maurya SP, Singh NP (2018) Comparing pre-and post-stack seismic inversion methods-a case study from Scotian Shelf, Canada. J Ind Geophys Union 22(6):585–597

Maurya SP, Singh KH (2019) Predicting porosity by multivariate regression and probabilistic neural network using model-based and colored inversion as external attributes: a quantitative comparison. J Geol Soc India 93(2):207–212

Maurya SP, Singh KH, Singh NP (2019) Qualitative and quantitative comparison of geostatistical techniques of porosity prediction from the seismic and logging data: a case study from the Blackfoot Field, Alberta, Canada. Mar Geophys Res 40(1):51–71

Maurya SP, Singh KH, Kumar A, Singh NP (2017) Reservoir characterization using post-stack seismic inversion techniques based on a real coded genetic algorithm. J Geophys 39(2):95–103

McCormack MD (1991) Neural computing in geophysics. Lead Edge 10(1):11–15

McCormack MD, Zaucha DE, Dushek DW (1993) First-break refraction event picking and seismic data-trace editing using neural networks. Geophysics 58(1):67–78

McCulloch WS, Pitts W (1943) A logical calculus of the ideas immanent in nervous activity. Bull Math Biophys 5(4):115–133

Murat ME, Rudman AJ (1992) Automated first arrival picking: a neural network approach 1. Geophys Prospect 40(6):587–604

Poulton MM, Sternberg BK, Glass CE (1992) Location of subsurface targets in geophysical data using neural networks. Geophysics 57(12):1534–1544

Pramanik AG et al (2004) Estimation of effective porosity using geostatistics and multi-attribute transforms a case study. SEG 69(2):352–372

Romeo G (1994) Seismic signals detection and classification using artificial neural networks. Annals of Geophysics 37:3

Röth G, Tarantola A (1994) Neural networks and inversion of seismic data. J Geophys Res Solid Earth 99(B4):6753–6768

Russell B, Hampson D, Schuelke J, Quirein J (1997) Multiattribute seismic analysis. Lead Edge 16(10):1439–1444

Russell SA, Reasoner C, Lay T, Revenaugh J (2001) Coexisting shear- and compressional-wave seismic velocity discontinuities beneath the central Pacific. Geophysical Research Letters 28(11):2281–2284

Schuelke JS, Quirein JA, Sag JF, Altany DA, Hunt PE (1997) Reservoir architecture and porosity distribution, Pegasus field, West Texas—an integrated sequence stratigraphic-seismic attribute study using neural networks. In SEG technical program expanded abstracts, society of exploration geophysicists, pp 668–671

Schultz PS, Ronen S, Hattori M, Corbett C (1994) Seismic-guided estimation of log properties (Part 1: a data-driven interpretation methodology). Lead Edge 13(5):305–310

Singh S, Kanli AI, Sevgen S (2016) A general approach for porosity estimation using artificial neural network method: a case study from Kansas gas field. Stud Geophys Geod 60(1):130–140

Soubotcheva N, Stewart RR (2004) Predicting porosity logs from seismic attributes using geostatistics. CREWES Research Report -Volume 1:1–4

Specht DF (1990) Probabilistic neural networks. Neural Netw 3(1):109–118

Specht DF (1991) A general regression neural network. IEEE Trans Neural Netw 2(6):568–576

Svozil D, Kvasnicka V, Pospichal J (1997) Introduction to multi-layer feed-forward neural networks. Chemometr Intell Lab Syst 39(1):43–62
Swisi, A.A., 2009. *Post-and Pre-stack attributes analysis and inversion of Blackfoot 3Dseismic dataset*, Doctoral dissertation, University of Saskatchewan
Taner MT (2001) Seismic attributes. CSEG recorder 26(7):49–56
Taner MT, Koehler F, Sheriff RE (1979) Complex seismic trace analysis. Geophysics 44(6):1041–1063
Taner MT, Schuelke JS, O'Doherty R, Baysal E (1994) Seismic attributes revisited. In: SEG technical program expanded abstracts, society of exploration geophysicists, pp 1104–1106
Taner MT, ODoherty R, Schuelke JS, Baysal E (1994) Seismic attributes revisited. Society of Exploration Geophysicists, Tulsa, OK (United States)
Todorov TI (2000) Integration of 3C-3D seismic data and well logs for rock property estimation. The University of Calgary
Tonn R (2002) Neural network seismic reservoir characterization in a heavy oil reservoir. Lead Edge 21(3):309–312
Torres-Verdin C, Victoria M, Merletti G, Pendrel J (1999) Trace-based and geostatistical inversion of 3-D seismic data for thin-sand delineation: an application in San Jorge Basin, Argentina. Lead Edge 18(9):1070–1077
Wang LX, Mendel JM (1992) Adaptive minimum prediction-error deconvolution and source wavelet estimation using Hopfield neural networks. Geophysics 57(5):670–679
Wu X, Ghaboussi J, Garrett JH Jr (1992) Use of neural networks in the detection of structural damage. Comput Struct 42(4):649–659
Zhang Y, Paulson KV (1997) Magnetotelluric inversion using regularized Hopfield neural networks. Geophys Prospect 45(5):725–743

The manufacturer's authorised representative in the EU is Springer Nature Customer Service Centre GmbH, Europaplatz 3, 69115 Heidelberg, Germany. If you have any concerns regarding our products, please contact ProductSafety@springernature.com

Printed and bound by CPI Group (UK) Ltd, Croydon, CR0 4YY
05/10/2026
02228339-0003